Springer Proceedings in Physics

Volume 269

Indexed by Scopus

The series Springer Proceedings in Physics, founded in 1984, is devoted to timely reports of state-of-the-art developments in physics and related sciences. Typically based on material presented at conferences, workshops and similar scientific meetings, volumes published in this series will constitute a comprehensive up-to-date source of reference on a field or subfield of relevance in contemporary physics. Proposals must include the following:

- name, place and date of the scientific meeting
- a link to the committees (local organization, international advisors etc.)
- scientific description of the meeting
- list of invited/plenary speakers
- an estimate of the planned proceedings book parameters (number of pages/ articles, requested number of bulk copies, submission deadline).

Please contact:

For Americas and Europe: Dr. Zachary Evenson; zachary.evenson@springer.com
For Asia, Australia and New Zealand: Dr. Loyola DSilva; loyola.dsilva@springer.com

More information about this series at http://www.springer.com/series/361

Ranjith G. Nair · Lalu Seban ·
Pujita Ningthoukhongjam
Editors

Proceedings of 28th National Conference on Condensed Matter Physics

Condensed Matter Days 2020 (CMDAYS20)

Springer

Editors
Ranjith G. Nair
National Institute of Technology Silchar
Silchar, India

Lalu Seban
National Institute of Technology Silchar
Silchar, India

Pujita Ningthoukhongjam
Department of Physics
National Institute of Technology Silchar
Silchar, India

ISSN 0930-8989 ISSN 1867-4941 (electronic)
Springer Proceedings in Physics
ISBN 978-981-16-5406-0 ISBN 978-981-16-5407-7 (eBook)
https://doi.org/10.1007/978-981-16-5407-7

This Springer imprint is published by the registered company Springer Nature Singapore Pte Ltd.
The registered company address is: 152 Beach Road, #21-01/04 Gateway East, Singapore 189721, Singapore

Preface

This book is a collection of the recent works of leading and upcoming researchers in India, highlighting the recent advances in the field of condensed matter physics. It features selected works presented in the 28th National Conference on Condensed Matter Physics, "Condensed Matter Days 2020 (CMDAYS20)", which was held during 11th to 13th December 2020. It was a virtual conference organized jointly by the Department of Physics and the Institute Innovation Cell of National Institute of Technology Silchar. Eminent researchers from all over the nation presented their works and ideas to an audience of their peers and young scientists. The conference also gave a platform to these young researchers to share their works in the field of condensed matter physics. Recent works and results in the fields of nanomaterials, biomaterials, magnetism, photocatalysis, photovoltaics, dielectrics, ferroelectrics, optoelectronics, soft condensed matter physics, thin films, and devices were presented during the event. The conference also featured a special session on "Hydrogen production, storage, sensing, and fuel cells".

Condensed matter physics is one of the most active and emerging fields of research. It involves both fundamental research and applied research. The field yields results that find application in various aspects of human life, from power generation to waste management, from memory storage to health sciences. India is a hotspot for emerging research in this field.

This book aims to highlight the latest research from the various areas of condensed matter physics that is happening in India. Researchers from various institutes and universities from all over the country have been featured in this book.

This book includes a total of twenty-four contributory papers with theoretical, simulation, and experimental results and three review papers. Through this book, the reader will get a clear idea of the current state of research in condensed matter physics and their practical application. This helps the readers to generate new ideas in their research and help in the implementation. We hope that this book will be of interest and help to new and old researchers alike.

Silchar, India

Ranjith G. Nair
Lalu Seban
Pujita Ningthoukhongjam

Contents

Contributors

Heremba Bailung Plasma Application Laboratory, Physical Sciences Division, Institute of Advanced Study in Science and Technology, Guwahati, India

Sabur A. Barbhuiya Department of Physics, Birla Institute of Technology and Science, Hyderabad Campus, Pilani, Hyderabad, India

Junmoni Barman Department of Physics, Rajiv Gandhi University, Arunachal Pradesh, India

Kashmiri Baruah Department of Physics, Tezpur University (Central University), Tezpur, India

Kshitij Bhargava Department of Electrical and Computer Science Engineering, Institute of Infrastructure, Technology, Research and Management, Ahmedabad, Gujarat, India

Vijay Bhatt Department of Physics, DDU College, University of Delhi, New Delhi, India

Aranya B. Bhattacherjee Department of Physics, Birla Institute of Technology and Science, Hyderabad Campus, Pilani, Hyderabad, India

P. K. Boruah Department of Instrumentation, Gauhati University, Guwahati, Assam, India

Palash Jyoti Boruah Plasma Application Laboratory, Physical Sciences Division, Institute of Advanced Study in Science and Technology, Guwahati, India

Hemen Ch. Medhi Department of Electronics St, Edmund's College, Shillong, Meghalaya, India

A. K. Chaudhary Advanced Centre of Research in High Energy Materials (ACRHEM), University of Hyderabad, Telangana, India

Avijit Chowdhury Organic Electronics and Sensor Laboratory, Department of Physics, National Institute of Technology Silchar, Silchar, Assam, India

Nipom Sekhar Das Organic Electronics and Sensor Laboratory, Department of Physics, National Institute of Technology Silchar, Silchar, Assam, India

Suma Das Organic Electronics & Sensor Laboratory, Department of Physics, National Institute of Technology Silchar, Cachar, Assam, India

Tanmay Das Department of Physics, Indian Institute of Technology Bhilai, GEC Campus, Sejbahar, Raipur, Chhattisgarh, India

Ujjal Das Department of Physics, National Institute of Technology, Silchar, Assam, India

Shivam Dave Department of Electrical and Computer Science Engineering, Institute of Infrastructure, Technology, Research and Management, Ahmedabad, Gujarat, India

P. Deb Department of Physics, Tezpur University (Central University), Tezpur, India

Rajesh Deb Department of Physics, National Institute of Technology Silchar, Silchar, Assam, India

Sulochana Deb Department of Physics, Gauhati University, Guwahati, India

Srikanta Debata Indian Institute of Technology Bhilai, GEC Campus, Sejbahar, Raipur, Chhattisgarh, India

A. V. Deshpande Visvesvaraya National Institute of Technology, Nagpur, India

V. K. Deshpande Department of Physics, Visvesvaraya National Institute of Technology, Nagpur, India

Jayanta Dey Department of Physics, Indian Institute of Technology Bhilai, GEC Campus, Sejbahar, Raipur, Chhattisgarh, India

A. Angeline Dorothy Hindustan Institute of Technology and Science, Chennai, India

D. Dwibedy Department of Physics, Veer Surendra Sai University of Technology, Burla, Sambalpur, India

Chandan Ghorui Advanced Centre of Research in High Energy Materials (ACRHEM), University of Hyderabad, Telangana, India

Sabyasachi Ghosh Department of Physics, Indian Institute of Technology Bhilai, GEC Campus, Sejbahar, Raipur, Chhattisgarh, India

Koustav Kashyap Gogoi Organic Electronics and Sensor Laboratory, Department of Physics, National Institute of Technology Silchar, Silchar, Assam, India

Piyush Jaiswal Department of Nanoscience and Technology, Centre for Advanced Studies, Dr. A.P.J. Abdul Kalam Technical University, Lucknow, India

Pradip K. Jha Department of Physics, DDU College, University of Delhi, New Delhi, India

Dipangkar Kalita Department of Physics, North-Eastern Hill University, Shillong, India

Rajesh Kumar National Institute of Technology, Silchar, India

Saumya R. Mohapatra Department of Physics, National Institute of Technology Silchar, Silchar, Assam, India

Manjula G. Nair Department of Physics, National Institute of Technology Silchar, Silchar, Assam, India

Ranjith G. Nair Solar Energy Materials Research & Testing Laboratory (SMaRT Lab), Department of Physics, National Institute of Technology Silchar, Silchar, Assam, India

P. Naveen Kumar Advanced Centre of Research in High Energy Materials (ACRHEM), University of Hyderabad, Telangana, India

S. K. Nikhil Solar Energy Materials Research & Testing Laboratory (SMaRT Lab), Department of Physics, National Institute of Technology Silchar, Silchar, Assam, India

Manas R. Panigrahi Department of Physics, Veer Surendra Sai University of Technology, Burla, India

Puspamitra Panigrahi Hindustan Institute of Technology and Science, Chennai, India

Santosh V. Patil Department of Electrical and Computer Science Engineering, Institute of Infrastructure, Technology, Research and Management, Ahmedabad, Gujarat, India

Debakeenandan Pradhan Department of Physics, Indian Institute of Technology Bhilai, GEC Campus, Sejbahar, Raipur, Chhattisgarh, India

G. M. Prasad Intelligent Mining Systems, CSIR-Central Institute of Mining and Fuel Research, Dhanbad, India

Koalla Rajesh Advanced Centre of Research in High Energy Materials (ACRHEM), University of Hyderabad, Telangana, India

S. Ravi Department of Physics, Indian Institute of Technology Guwahati, Guwahati, India

Asim Roy Department of Physics, National Institute of Technology, Silchar, Assam, India

A. K. Sahoo Department of Physics, Veer Surendra Sai University of Technology, Burla, Sambalpur, India

Megha A. Salorkar Department of Physics, Visvesvaraya National Institute of Technology, Nagpur, India

Jaysri Sarkar Department of Physics, Gauhati University, Guwahati, India

Peddy Satyanarayana Visvesvaraya National Institute of Technology, Nagpur, India

Atul Saxena Department of Physics, North-Eastern Hill University, Shillong, India

Joshi Pushkar Shrikant Solar Energy Materials Research & Testing Laboratory (SMaRT Lab), Department of Physics, National Institute of Technology Silchar, Silchar, Assam, India

Dhruv Pratap Singh Indian Institute of Technology Bhilai, GEC Campus, Sejbahar, Raipur, Chhattisgarh, India

J. K. Singh Intelligent Mining Systems, CSIR-Central Institute of Mining and Fuel Research, Dhanbad, India

Narendra Singh Department of Nanoscience and Technology, Centre for Advanced Studies, Dr. A.P.J. Abdul Kalam Technical University, Lucknow, India;
Department of Chemical Engineering, Indian Institute of Technology Tirupati, Tirupati, Andhra Pradesh, India

Meenakshi Srivastava Department of Nanoscience and Technology, Centre for Advanced Studies, Dr. A.P.J. Abdul Kalam Technical University, Lucknow, India

F. A. Talukdar Department of Electronic and Communication Engineering, National Institute of Technology Silchar, Silchar, India

Anita Tamang Department of Physics, Indian Institute of Technology Bhilai, GEC Campus, Sejbahar, Raipur, Chhattisgarh, India

S. K. Tripathy Department of Electronic and Communication Engineering, National Institute of Technology Silchar, Silchar, India

Tohru Tsuruoka International Center for Materials Nanoarchitectonics (WPI-MANA), National Institute for Materials Science, Tsukuba, Japan

Sesha Vempati Department of Physics, Indian Institute of Technology Bhilai, GEC Campus, Sejbahar, Raipur, Chhattisgarh, India

Chapter 1
Molybdenum Disulphide Nanosheet: Hydrothermal Synthesis and Characterization

Sulochana Deb and Jaysri Sarkar

Abstract In the field of nanotechnology, MoS_2 nanosheets have a wide range of applications because of their unique properties. Here, we report hydrothermal method of synthesis of MoS_2 nanosheet. The as- synthesized MoS_2 nanosheets are characterized by the Scanning electron spectroscopy (SEM), X ray diffraction (XRD) studies, UV visible spectroscopy, Photoluminescence (PL) spectroscopy and I-V characteristics to determine their morphological, structural, optical and electrical properties. The SEM image confirms the synthesis of well-shaped nanosheet like structure with length 9.6 μm and breadth 3.2 μm but the thickness is in the nanorange. The UV–Vis spectra show the absorption peak of MoS_2 at 313.13 nm and a band gap calculated is found to be 2.68 eV. The XRD peaks are observed at 14°, 25.46° and 33.79° corresponding to (002), (111) and (100) plane. XRD also gives the crystalline size of nanoparticles about 112.8 nm. Photoluminescence spectra shows that the maximum emission peak is at 465.79 nm for excitation wavelength at 350 nm. The I-V characteristic gives the information that the curve is nonlinear, asymmetric and the p–n junction are formed with the high reverse break-down voltage of MoS_2.

1.1 Introduction

Two-dimensional materials have attracted the most attention due to their potential applications in different fields [1–3]. They exhibit completely different properties compared to their bulk counterparts for new generations of electronic and optoelectronic devices [4–6]. Transition metal dichalcogenides (TMDs) such as MoS_2, WS_2, WTe_2, TiS_2, TaS_2, ZrS_2, or graphite are important two-dimensional (2D) materials and the structures of these layered materials are hexagonally packed. Unlike graphene which has no bandgap, the TMDs are natural semiconductors having non-zero band gap. Therefore, they have wide applications in electronic devices and optical devices. MoS_2 nanosheet exhibit unique physiochemical, biological, mechanical, optical and

S. Deb (✉) · J. Sarkar
Department of Physics, Gauhati University, Guwahati 781014, India
e-mail: debsulochana@gauhati.ac.in

R. G. Nair et al. (eds.), *Proceedings of 28th National Conference on Condensed Matter Physics*, Springer Proceedings in Physics 269,
https://doi.org/10.1007/978-981-16-5407-7_1

electrical properties [7, 8]. It has good chemical and thermal stability also. The electronic properties and the energy gap of MoS_2, are strongly related to the number of atomic layers. Few-layered MoS_2 turns into a direct bandgap semiconductor if thinned whereas the bulk MoS_2 has an indirect bandgap, which is favourable for optoelectronics [9–11]. Due to these distinctive properties, MoS_2 can be used in numerous fields such as biomedicine, energy storage, gas sensing, catalysis, hydrogen evolution and electronics engineering [12, 13]. Consequently, finding an effective method to prepare uniform-size monolayer MoS_2 nanosheets seems essential. The literature review shows that the current research on two-dimension (2D) MoS_2 nanosheets is principally based on the exfoliation technique, chemical vapor deposition method, or solvothermal route [14–16].

In this study, we report simple hydrothermal method of preparation and characterization of wide band gap MoS_2 nanosheets. The Wide bandgap (WBG) semiconductors such as SiC, GaN, and diamond have bandgaps in the range 2–6 eV [17, 18]. The wide band gap semiconductors can be operated at much higher temperatures, voltages, and frequencies [19]. Therefore, they are suitable to use in optoelectronic and electronic devices.

1.2 Experimental Details

1.2.1 Materials

Materials are acquired from Merck Company and 99.9% pure. To prepare the MoS_2 nanosheet, Ammonium heptamolybdate tetrahydrate $(NH_4)_6Mo_7O_{24}4H_2O$, Citric acid $C_6H_8O_7$ and Thiourea CH_4N_2S, are used. Deionized water is used as medium although.

1.2.2 Method

At first 2 g of ammonium heptamolybdate tetrahydrate and 0.7 g citric acid were dissolved in 20 ml distilled water and kept at 90 °C with continuous stirring for 30 min. Then 1.3 g of thiourea is dissolved in 10 ml distilled water and then added drop wise to the above solution. The pH of the solution is maintained at 4.28. The colour of the solution changes from colourless to blue. Finally, the prepared dispersion is transferred into a Teflon autoclave and kept for 2 h. at 150 °C. After that we see that the colour of the solution change to dark blue. The schematic diagram for synthesis of MoS_2 nanosheet is shown in Fig. 1.1.

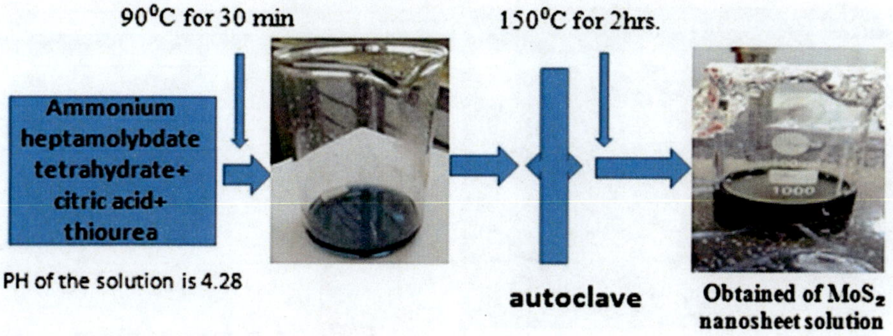

Fig. 1.1 Schematic diagram for hydrothermal synthesis of MoS_2 nanosheet

1.2.3 Characterization

To study the morphology of the as prepared MoS_2 nanosheets sample, Field Emission Scanning Electron microscopy (FESEM) imaged are collected by FESEM (ZEISS). Stuctural properties are characterized by X-Ray diffraction (XRD) data collected by Scifert XRD 3000 pd Diffractometer with Cu-Kα (0.15418 nm) radiation. For optical properties, UV–Vis optical absorption spectra is recorded by CARY-300 spectrophotometer and Photoluminescence (PL) spectra is taken by JASCO FP- 8300 Spectroflurometer. Electrical properties of the nanosheet films are measured through recording data for current–voltage (I-V) curves by measuring current using keithly 2400 source meter.

1.3 Results and Discussions

1.3.1 Morphology

The morphological study of as prepared MoS_2 nanosheet is done with the help of Field emission scanning electron microscope (FESEM). Figure 1.2a shows the FESEM image of the sample before heating in the autoclave. We can see that the sample is not exactly sheet like structure. Figure 1.3a Shows FESEM image of MoS_2 nanosheet after heating at 150 °C for 2 h. The image shows that well shaped nanosheets are formed with approximate length 9.662 μm and breadth 3.25 μm. Figures 1.2b and 1.3b show the EDX of the sample before and after heating respectively.

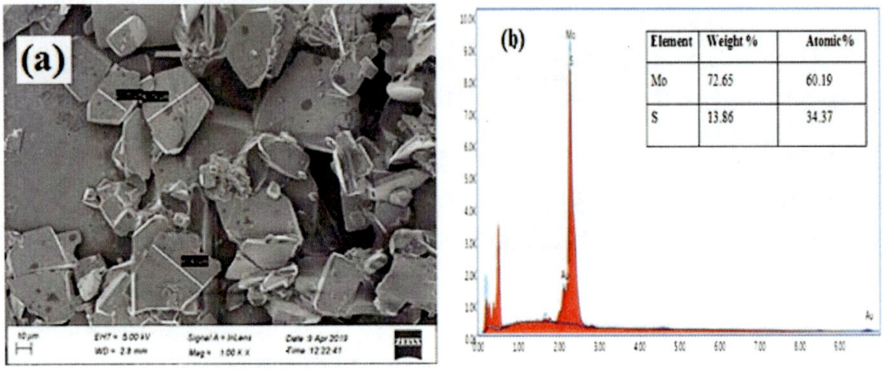

Fig. 1.2 **a** FESEM image **b** EDX of the of MoS_2 nanocomposite before heating

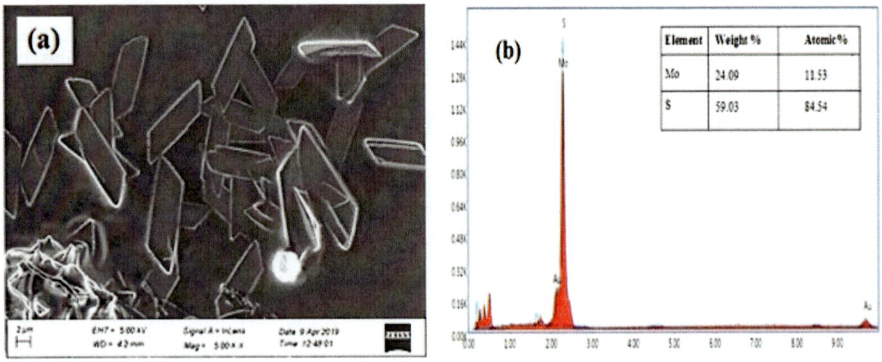

Fig. 1.3 **a** FESEM image of MoS_2 nanosheet **b** EDX of MoS_2 nanosheet

1.3.2 Structure

XRD pattern of the nanocomposite films are shown in Fig. 1.4. The characteristic diffraction peaks at 2θ values of 14°, 25.46°, 33.79°, and 60° corresponding to the (002), (111), (100) and (110) crystal plane of the MoS_2 structure. The broad nature of the XRD peak could be attributed to formation of nanosized particles. From conventional Scherrer formula given below the crystalline size (D) can be calculated:

$$D = k\lambda/\beta_{1/2}\cos\theta$$

where β ½ is full- width at half maximum of the peak at 2Θ, k is constant (k = 0.9) and λ = 0.154 nm is the CuKα1 wavelength. The average crystalline size is found to be 112 nm.

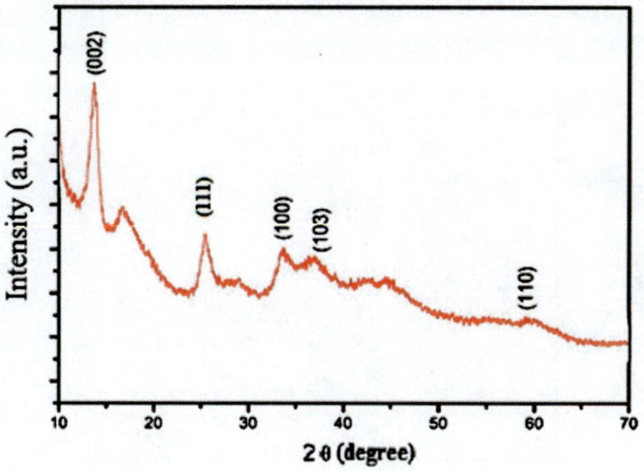

Fig. 1.4 XRD of MoS_2 nanosheet

1.3.3 Optical Properties

1.3.3.1 UV–Visible Spectra and Band Gap Calculation

The UV–Visible absorption spectra is shown in Fig. 1.5a. The absorption peak of MoS_2 nanosheet is found at 31 3.13 nm. The indirect band gap of the Molybdenum Disulfide nanosheet has been determined by from the intercepts of the conventional Tauc relation $(\alpha h\upsilon)^{½} = A\,(h\upsilon - E_g)$,where α is absorbance, hυ is the incident photon energy and E_g is the band gap of the semiconductor. These plots are shown in Fig. 1.5. Band gap for indirect transition may be determined from the extrapolation of the linear section of the curves to x-axis for $(\alpha(v)hv)^{1/2} = 0$. The indirect band gap of MoS_2 nanosheet calculated from UV–Visible absorption spectra by "Tauc plot" is found to be at 2.68 eV. The calculated band gap energy is greater than that of monolayer MoS_2 (1.9 eV) and bulk MoS_2 (1.2 eV) [10, 20, 21] but less than that of

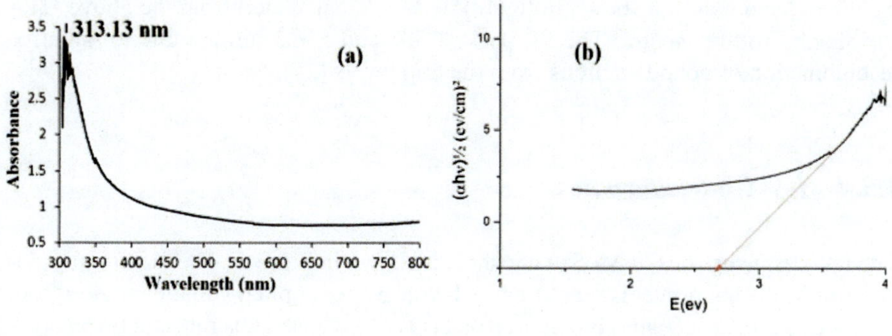

Fig. 1.5 **a** UV–visible spectra of MOS_2 nanosheet **b** Band gap for the MOS_2 nanosheet

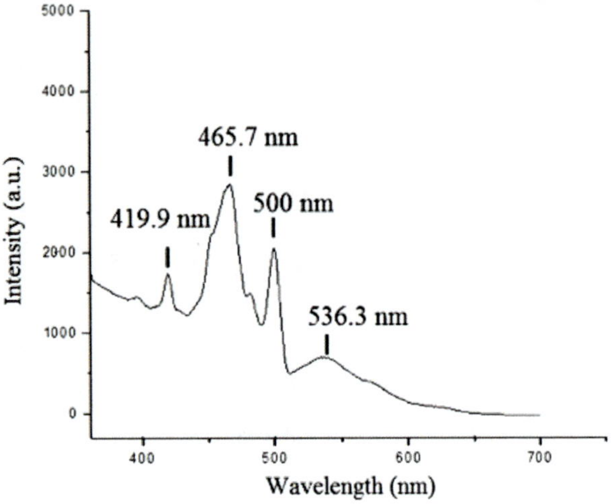

Fig. 1.6 PL spectra of MoS_2 nanosheet

monolayer MoS_2 quantum dot (4.96 eV) [22] and MoS_2 nanosheets (4.7 eV) [23]. From the calculated band gap it can be inferred that the prepared MoS_2 nanosheets fall in the category of wide band gap semiconductors [17, 18].

1.3.3.2 Photoluminescence Spectra

The room temperature photoluminescence (PL) spectra of the MoS_2 nanoparticles is shown in Fig. 1.6. From PL spectra of MoS_2 nanosheet, it has been observed that the emission peak of maximum intensity of emitted light is at 465.7 nm for excitation wavelength of 350 nm. Band gap calculated from this PL peak is found to be 2.66 eV which is slightly less than the indirect band gap calculated from UV–Visible absorption spectra. The PL peak at 419.19 nm (2.95 eV) is caused by direct band gap transition at the K point of the Brillion zone [24] which is very close to the direct band gap (not shown in figure) of 3.1 eV calculated from the above UV–Visible absorption spectra. The PL peak at 500 and 536.3 nm are due to radiative recombination of bound excitons from the trap levels [25].

1.3.4 I-V Characteristics

The I-V characteristics of MoS_2 nanosheet is shown in Fig. 1.7. The electrical properties of MoS_2 nanosheet is measured by I-V curve for applied voltage ranging from -20 V to $+20$ V. MoS_2 film is prepared on ITO coated glass substrate and I-V characteristics is measured using two probe method and perpendicular configuration. On the

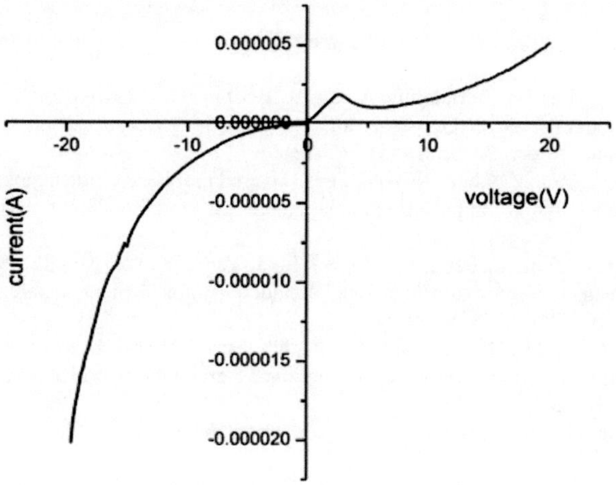

Fig. 1.7 I-V characteristics of MoS_2 nanosheet

top of the MoS_2 film small area is coated with silver paint and connections are taken by fine copper wires. I-V characteristics of MoS_2 clearly demonstrate a nonlinear and asymmetric I-V curve and also the p–n junction is formed by the extremely high reverse break-down voltage.

1.4 Conclusions

We have successfully synthesized MoS_2 nanosheet of wide band gap by simple hydrothermal method. SEM images reveal the formation of MoS_2 nanosheet with length 9.662 μm and breadth 3.25 μm. XRD analysis exhibits the crystalline nature with three different peaks at 14°, 25.46° and 33.79° corresponding to plane (002), (111) and(100) of the crystal. From UV–visible spectra absorption peak is observed at 313.13 nm. Indirect band gap calculated from UV–visible spectra is 2.68 eV. The emission peak of MoS_2 nanosheet is found to be at 465.7 nm for excitation wavelength 350 nm. Band gap calculated from PL peak is found to be at 2.66 eV. From I-V characteristic of MoS_2 we clearly demonstrate a nonlinear and asymmetric I-V curve and also the p–n junction is formed.

Acknowledgements The authors express their deep sense of gratitude to SAIF, Gauhati University, Guwahati for SEM and XRD measurement, Dr. Manas P. C. Kalita and Prof. D. Sarkar, Dept. of Physics, Guwahati for PL and electrical measurements respectively.

References

1. P. Néstor, A.L. Elías, A. Berkdemir, A. Castro-Beltran, H.R. Gutiérrez, S. Feng, R. Lv, T. Hayashi, F. López-Urías, S. Ghosh, et al., Photosensor device based on few-layered WS2 films. Adv. Funct. Mater. **23**, 5511–5517 (2013)
2. G. Zhao, J. Li, X. Ren, C. Chen, X. Wang, Few-layered graphene oxide nanosheets as superior sorbents for heavy metal ion pollution management. Environ. Sci. Technol. **45**, 10454–10462 (2011)
3. F. Jun, X. Sun, C. Wu, L. Peng, C. Lin, S. Hu, J. Yang, Y. Xie, Metallic few-layered VS2 ultrathin nanosheets: high two-dimensional conductivity for in-plane supercapacitors. J. Am. Chem. Soc. **133**, 17832–17838 (2011)
4. M. Jafari, M. Asadpour, N.A. Majelan, M. Faghihnasiri, Effect of boron and nitrogen doping on electro-optical properties of armchair and zigzag graphyne nanoribbons. Comput. Mater. Sci. **82**, 391–398 (2014)
5. Q. Ke, J. Wang, Graphene-based materials for supercapacitor electrodes—a review. J. Mater. **2**, 37–54 (2016)
6. B. Francesco, A. Bartolotta, J.N. Coleman, C. Backes, 2D-crystal-based functional inks. Adv. Mater. **28**, 6136–6166 (2016)
7. V. Gunasekarana, M.K. Devarajub, S. Yuvarajc, V.J.Y. Suryac, Varu Singhd, K. Karthikeyane, S.-J. Kime, Electrical transport properties of two-dimensional MoS2 nanosheets synthesized by novel method. Mater. Sci. Semicond. Process. **66**, 81–86 (2017)
8. O. Lopez-Sanchez, D. Lembke, M. Kayci, A. Radenovic, A. Kis, Ultrasensitive photodetectors based on monolayer MoS_2. Nat. Nanotechnol. **8**, 497–501 (2013)
9. P. Erwin, L. Gelato, B. Chabot, M. Penzo, K. Cenzual, R. Gladyshevskii, *TYPIX Standardized Data and Crystal Chemical Characterization of Inorganic Structure Types* (Springer, Berlin/Heidelberg, Germany, 2013)
10. S. Lebegue, O. Eriksson, Electronic structure of two-dimensional crystals from ab initio theory. Phys. Rev. B **79**, 115409 (2009).
11. M.K. Fai, C. Lee, J. Hone, J. Shan, T.F. Heinz, Atomically thin MoS_2: a new direct-gap semiconductor. Phys. Rev. Lett. **105**, 136805 (2010)
12. Y. Shi, W. Zhou, A. Lu, W. Fang, Y. Lee, A.L. Hsu, S.M. Kim, K.K. Kim, H.Y. Yang, L. Li et al., van der Waals epitaxy of MoS_2 layers using graphene as growth templates. Nano Lett. **12**, 2784–2791 (2012)
13. B. John, M. Li, S. Wang, P. Wang, P. Papakonstantinou, Electrocatalytic hydrogen evolution reaction on edges of a few layer molybdenum disulfide nanodots. ACS Appl. Mater. Interfaces **7**, 14113–14122 (2015)
14. E. Benavente, M.A.S. Ana, F. Mendizábal, G. González, Intercalation chemistry of molybdenum disulfide. Coord. Chem. Rev. **224**, 87–109 (2002)
15. Y.S. Lee, X.Q. Zhang, W. Zhang, M.T. Chang, C.T. Lin, K.D. Chang, Y.C. Yu, J.T. Wang, C.S. Chang, L.J. Li, T.W. Lin, Synthesis of large-area MoS_2 atomic layers with chemical vapor deposition. Adv. Mater. **24**, 2320–2325 (2012)
16. J. Xie, H. Zhang, S. Li, R. Wang, X. Sun, M. Zhou, J. Zhou, X.W. Lou, Y. Xie, Defect-rich MoS_2 ultrathin nanosheets with additional active edge sites for enhanced electrocatalytic hydrogen evolution. Adv. Mater. **25**, 5807–5813 (2013)
17. G. Prestopino, M. Marinelli, E. Milani, C. Verona, G. Verona-Rinati, Transient lateral photovoltaic effect in synthetic single crystal diamond. Appl. Phys. Lett. **111**, 143504 (2017)
18. E. Monroy, F. Omnès, F. Calle, Wide-bandgap semiconductor ultraviolet photodetectors. Semicond. Sci. Technol. **18**, R33–R51 (2003)
19. E. Ahmed, T.P. Chow, Silicon carbide benefits and advantages for power electronics circuits and systems. Proc. IEEE **90**, 969–986 (2002)
20. Š Václav, J. Henych, Strongly luminescent monolayered MoS_2 prepared by elective ultrasound exfoliation. Nanoscale **5**, 3387–3394 (2013)

21. H. Hao, C. Du, H. Shi, X. Feng, J. Li, Y. Tan, W. Song, Water-soluble monolayer molybdenum disulfide quantum dots with upconversion fluorescence. Part. Part. Syst. Charact. **32**, 72–79 (2015)
22. X. Ren, L. Pang, Y. Zhang, X. Ren, H. Fan, S.F. Liu, One-step hydrothermal synthesis of monolayer MoS_2 quantum dots for highly efficient electrocatalytic hydrogen evolution. J. Mater. Chem. A **3**, 10693–10697 (2015)
23. M. Mitra, K. Salimeh, A. Fahimeh, Preparation of few-layered wide bandgap MoS_2 with nanometer lateral dimensions by applying laser irradiation. Curr. Comput.-Aided Drug Des. **10**(3), 164 (2020)
24. M.K. Fai, C. Lee, J. Hone, J. Shan, T.F. Heinz, Atomically thin MoS_2: a new direct-gap semiconductor. Phys. Rev. Lett. **105**, 136805(4) (2010)
25. P.V. Hung, K. Kim, D. Jung, K. Singh, E. Oh, Chung, J.S. Liquid phase co-exfoliated MoS_2–graphene composites as anode materials for lithium ion batteries. J. Power Sources **244**, 280–286 (2013)

Chapter 2
Liquid Phase Exfoliation and Microwave Assisted Modification in MoS_2 Nanostructure

Rajesh Deb, Rajesh Kumar, Manjula G. Nair, and Saumya R. Mohapatra

Abstract In the present study, we develop one-dimensional nanostructures of MoS_2 from its two-dimensional (2D) nanosheets by following self-assembling process in the presence of microwave radiation. First, we exfoliated MoS_2 into 2D nanosheets by liquid phase exfoliation method. Then the solutions of MoS_2 nanosheets are placed in a microwave oven. By varying the exposure time of microwave (IR) radiation on MoS_2 solution, we investigated its effect on the micro-structural evolution of MoS_2 nanosheets. From the SEM images, it is observed that with an exposure time of 2.5 min, nanorods of MoS_2 start emerging as if piercing through the nanosheets of MoS_2. At 5 min exposure span, the nanorods start growing further making a network like structure. But the transformation from nanosheets to nanorods was not complete. Still some nanosheets of MoS_2 were visible. At 7.5 min exposure time, the nanorods are grown in diameter as well as in length and with no appearance of nanosheets. The 2D nanosheets either could have completely transformed into nanorods or partially may self- assembled into bulk structures. This study suggests that microwave irradiation is a facile way to develop one-dimensional nanostructures of MoS_2 from two-dimensional nanosheets.

2.1 Introduction

Transition metal dichalcogenides (TMDCs) in the form of atomically thin monolayer or few-layer nanosheets are very popular two-dimensional (2D) materials with extensive research in the areas of nanoelectronics [1], energy storage [2], photocatalysis [3] and sensor [4] applications. Just like the 2D form of TMDCs, zero-dimensional form (quantum dots) and one-dimensional form (nanowires or nanotubes) of TMDCs are also equally drawing the enthusiasm among the researchers community. For these three classes of TMDC nanomaterials, the preparation route may be broadly similar in terms of either bottom-up or top-down approaches. Particularly, the one-dimensional

R. Deb · R. Kumar · M. G. Nair · S. R. Mohapatra (✉)
National Institute of Technology, Silchar 788010, India
e-mail: saumya@phy.nits.ac.in

R. G. Nair et al. (eds.), *Proceedings of 28th National Conference on Condensed Matter Physics*, Springer Proceedings in Physics 269,
https://doi.org/10.1007/978-981-16-5407-7_2

nanostructures of TMDCs in the form of nanowires, nanotubes, nanorods or nanoscrolls are mostly prepared from bottom-up approaches using hydrothermal methods [5–7]. However, there is also another possibility to develop one-dimensional nanostructures from the 2D nanosheets due to self-assembly process. This method is less cumbersome and also the one-dimensional nanostructures prepared from 2D nanosheets have all the good properties of nanosheets and show new functionalities due to change in aspect ratio [8].

Here, we prepared 2D nanosheets of MoS_2 from bulk MoS_2 by liquid phase exfoliation method. Then by exposing the nanosheets of MoS_2 to microwave radiation, one-dimensional nanostructures such as MoS_2 nanorods were prepared and further characterized.

2.2 Experimental Methods

Bulk MoS_2 powder was purchased from Alfa Aesar. DMF was chosen as the medium of exfoliation and was purchased from Merck. All the chemicals were used as received without any further modification. Firstly, 500 mg of bulk MoS_2 powder were grinded for about half an hour by adding few drops of DMF in regular short intervals and the paste was further diluted by adding 300 ml of DMF. The solution was then subjected to probe sonication using a probe sonicator at 100 W power for 4 h and it was done with 12 s on and 8 s off pulse cycles. The resulting solution was centrifuged at 2000 rpm for 5 min and again at 3000 rpm for 5 min to separate the exfoliated MoS_2 nanosheets from the bigger MoS_2 particles [9]. They are further used for modification under microwave irradiation.

2.2.1 Calculation of Yield of Exfoliated MoS_2

The exfoliated MoS_2 shows good stability and no settlement or re-stacking of the nano-sheets of MoS_2 was observed even after storage of one month as shown in Fig. 2.1. Also, no change in color was observed in the storing bottles. To calculate the yield of exfoliated MoS_2, 5 ml of exfoliated MoS_2 was taken in the clean glass petridish and it was dried at 180^0C for 6 h on a hot plate to remove the DMF solvents. The yield of MoS_2 is the difference between the weight of petridish with dry MoS_2 to that of the empty petridish. The yield of MoS_2 was calculated to be 12 mg/10 ml as shown in Table 2.1.

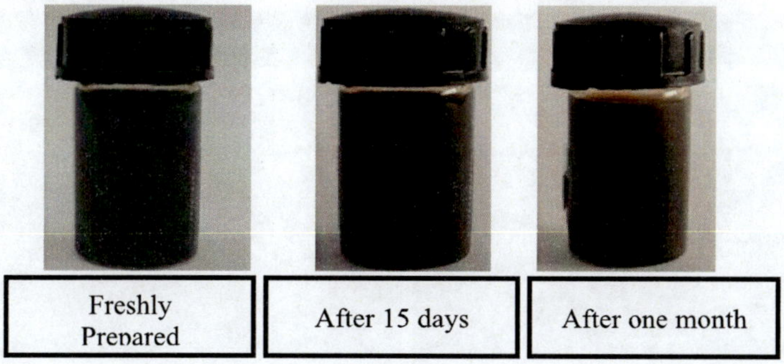

Fig. 2.1 Picture of exfoliated MoS_2 dispersed in DMF taken at different intervals of time

Table 2.1 Calculation of yield of exfoliated MoS_2

Weight of petridish (a)	Weight of petridish + dried MoS_2 (b)	Weight of MoS_2 ($b-a$)	Yield of exfoliation
41.2163 gm	41.2223 gm	0.006 gm	12 mg/10 ml

2.2.2 *Microwave Irradiation on Exfoliated MoS_2*

The exfoliated MoS_2 dispersed in DMF were taken in four clean glass bottles. Each bottle contains 10 ml of the solution and put into microwave oven separately for 0, 2.5, 5 and 7.5 min. The power rating of the microwave oven was fixed at 750 W. The microwave treated solutions were then used for characterization. Structural, micro-structural and optical characterizations were carried out using X-ray diffraction (XRD), scanning electron microscope (SEM), UV–Vis spectroscopy and photoluminescence spectroscopy.

2.3 Results and Discussions

2.3.1 *Morphological Characterization*

One of the important characterizations of the MoS_2 nanostructures after microwave irradiation is the scanning electron microscopy (SEM) which will give the direct evidence of any change that happened due to microwave irradiation. Figure 2.2 shows the SEM images of exfoliated MoS_2 samples exposed to microwave for 0, 2.5, 5.0 and 7.5 min. These SEM images were taken for the exfoliated samples deposited on a glass substrate after exposing to microwave. Figure 2.2a clearly shows the nanoplatelets of exfoliated MoS_2 self-assembled into random clusters on the surface of the glass plate as it was dried by drop casting of the MoS_2 solutions. At 2.5 min microwave

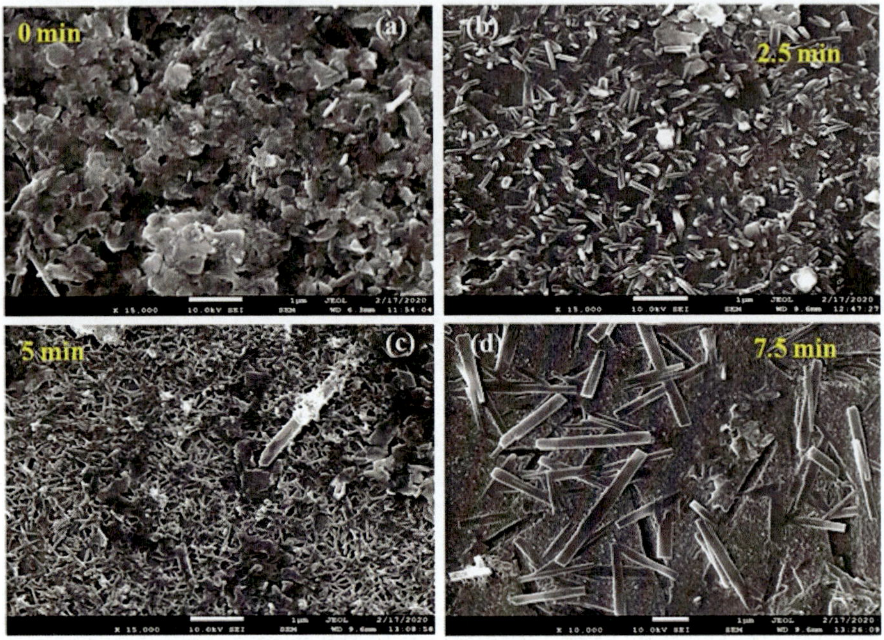

Fig. 2.2 SEM images of MoS_2 nano-structures with microwave exposure time **a** 0 min, **b** 2.5 min, **c** 5 min, and **d** 7.5 min

exposure time, small nanorods start emerging from the platelets as if piercing through as shown in Fig. 2.2b. The length of the nanorods is quite small, in a range of 0.3–0.5 μm. The average diameter of the nanorods is less than 100 nm. However, the transformation of two-dimensional nanosheets to one-dimensional nanorods was not complete in 2.5 min of microwave exposure time. That's why both the nanostructures were visible in the sample. After further increasing the exposure time to 5 min, the microstructure further changes and the nanorods started further growing. They made a network like structure of nanorods with emergence of few bigger nanorods. As shown in Fig. 2.2d, the nanorods looks quite big with length of 3–5 μm which is 10 times larger than MoS_2 nanorods prepared with 2.5 min exposure time. Similarly diameter also increased to ~300 nm. The nanoplatelets are now not visible, though in the background it seems that nanoplatelets those didn't transform to nanorods, made bigger clusters or bulk MoS_2. These results suggest that microwave irradiation up to 7.5 min is enough to make nanorods from the nanoplatelets of MoS_2.

2.3.2 *Structural Characterization*

Structural characterization of the MoS_2 samples was carried out using X-ray diffraction (XRD). The XRD pattern of four samples was plotted in the 2θ range of 10°–60°

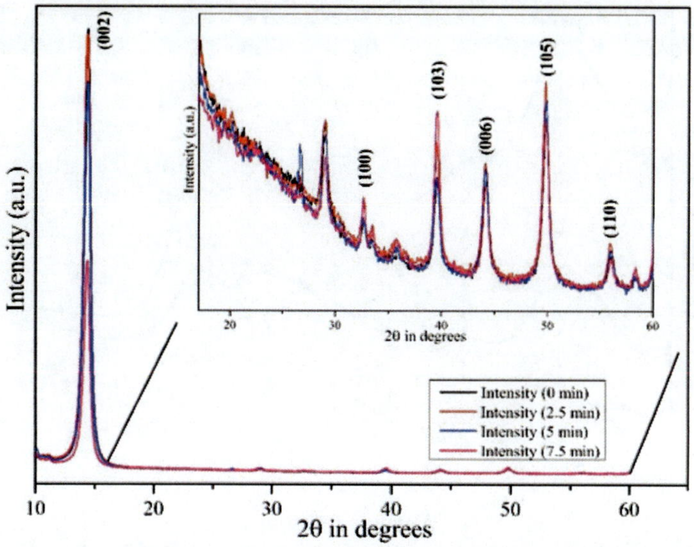

Fig. 2.3 XRD pattern of nano-structured MoS_2 with different microwave exposure time

and presented in the Fig. 2.3. The most intense Bragg peak observed at $2\theta = 14.6°$ is due to stacking along *C*-axis i.e. (002) peak. Other Bragg peaks are also observed for the MoS_2 samples as shown in the inset for 2θ in the range of 20°–60°. It is observed that with increasing the exposure time to microwave radiation, the intensity of the (002) Bragg peaks is decreasing. This suggests that the growth of the MoS_2 nanorods from the nano-platelets may not be favored along the *C*-axis. This trend of decreasing intensity with increasing microwave exposure time is not a general trend for other Bragg peaks. As an example, for the Bragg peak (103) observed at 39.6°, the inverse is happening i.e. with increasing exposure time to microwave, the peak intensity is increasing. This may be due to the increase in lateral size of the nanorods with microwave exposure time.

2.3.3 *Optical Characterization*

The optical characterization of the exfoliated MoS_2 nanosheets and microwave irradiated samples were performed in the solution form using UV–Vis and photoluminescence spectroscopy. Figure 2.4 shows the UV–visible spectra of the samples with microwave irradiation period of 0, 2.5, 5.0, and 7.5 min. The major absorbance peaks appearing for the sample with no exposure to microwave radiation (0 min) are 656, 595, 424 and 348 nm. The peaks at 656 and 595 nm which are referred to as excitons A and B, respectively, were results of the direct transition at the K-pop point in the Brillouin Zone [10]. There is an energy separation of 0.18 eV between A and B

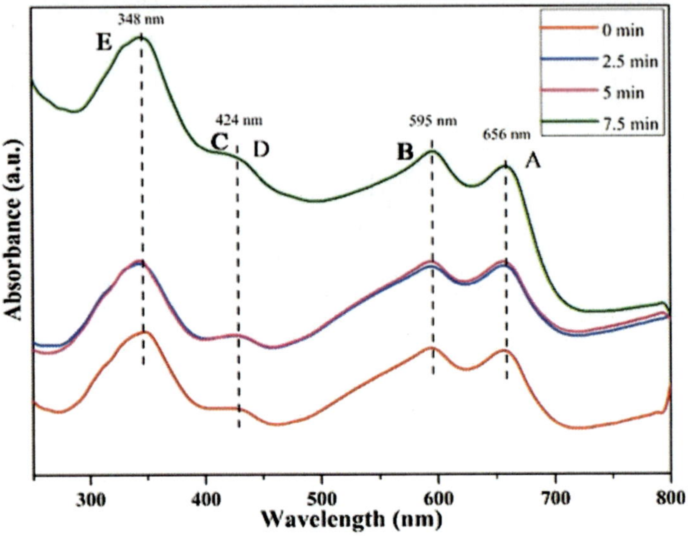

Fig. 2.4 UV–Vis absorption spectra of MoS_2 nano-structures

exactions which is due to the spin–orbit splitting of the valence band at the K-point [11]. The C, D (broad peak at ~424 nm) and E (348 nm) were attributed to the direct transition of the deep valance band to conduction band [12–14]. The shape of the spectra and relative intensity of these excitons mostly depend upon layer thickness and also lateral size of the MoS_2 clusters. It is observed that the relative intensity of C, D and E peaks increases with respect to the A and B excitons with increase in microwave exposure time.

The photoluminescence (PL) spectra of the MoS_2 samples are presented in Fig. 2.5. Two different excitation wavelengths were chosen at 343 and 595 nm to study the low energy excitons like C, D and E; high energy excitons like A and B separately. They are given in Fig. 2.5a and b, respectively. In both the figures, it is observed that all the excitons are red-shifted on microwave irradiation. The red-shift amount also increases with increase in irradiation time.

Earlier reports suggest that in case of quantum dots of MoS_2, the PL spectra shows blue shift in comparison to bulk MoS_2 samples [15, 16]. In our present work, due to microwave irradiation assisted self-assembling of MoS_2 nano-platelets, there is a change in aspect-ratio of the MoS_2 nano-structures. As the nanorods are getting bigger and thicker with increasing irradiation time, the red-shift is also becoming more prominent as observed for 7.5 min exposure time.

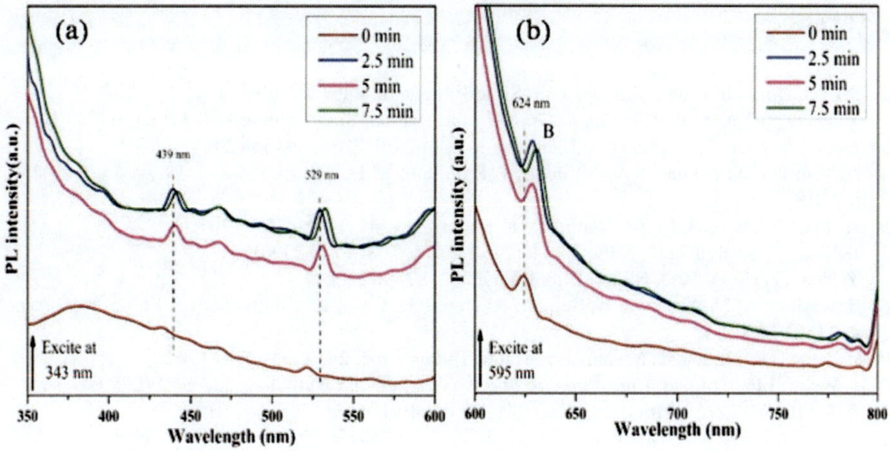

Fig. 2.5 Photoluminescence spectra of MoS_2 nanostructure samples with different exposure time to microwave; the excitation wavelength for **a** and **b** are 343 nm and 595 nm, respectively

2.4 Conclusion

In summary, one-dimensional nanostructures like MoS_2 nanorods were synthesized by liquid phase exfoliation method and further modified them by exposing to microwave radiation. The samples were exposed to microwave radiations for different intervals of time using a standard household microwave oven. The SEM images gave direct evidence of transformation of two-dimensional nanosheets of MoS_2 to one-dimensional MoS_2 nanorods. With increasing exposure time, nanorod size (length as well as lateral) increased. At 7.5 min of exposure time to microwave, most of the nano-platelets get converted to nanorods with some are self-assembled to get bulk size back. The structural investigation using XRD shows the intensity of characteristic Bragg peak at 14.6° due to (002) planes is decreasing with increase in microwave exposure time. This suggested that the nanorod growth is not along the c-axis. The optical characterization using UV–Vis and PL spectroscopy also shows systematic change in the sample spectra in terms of peak intensity and position on variation of exposure time to microwave. These studies confirm that exposing microwave is a facile way to prepare one-dimensional nano-structures like nanorod and nanotube of MoS_2 from the exfoliated nano-sheets. These one dimensional nano-structures of MoS_2 can have similar and specific application potentials in comparison to their conventional two-dimensional counterparts.

Acknowledgements The authors greatly acknowledge Centre for Nanotechnology, IIT Guwahati for providing SEM instrumentation facility.

References

1. S.J. McDonnell, R.M. Wallace, Thin Solid Films **616**, 482–501 (2016)
2. X. Duan, J. Xu, Z. Wei, J. Ma, S. Guo, H. Liu, S. Dou, Small Methods **1**, 1700156 (2017)
3. Z. Lia, X. Menga, Z. Zhang, J. Photochem. Photobiol. C **35**, 39–55 (2018)
4. N. Ashrafa, M.I. khana, A. Majida, M. Rafiqueb, M.B. Tahira, Chin. J. Phys. **66**, 246–257 (2020)
5. H. Lin, X. Chen, H. Li, M. Yang, Y. Qi, Mater. Lett. **64**, 1748–1750 (2010)
6. J.R. Ota, S.K. Suneel, J. Nanosci. Nanotechnol. **6**, 168–174 (2006)
7. Y. Tian, Y. He, Y. Zhu, Mater. Chem. Phys. **87**, 87–90 (2004)
8. S. Reshmi, M.V. Akshaya, B. Satpati, A. Roy, P.K. Basu, K. Bhattacharjee, Mater. Res. Exp. **4**, 115012 (2017)
9. P. Pathak, R. Deb, S.R. Mohapatra, Mater. Today: Proc. **24**, 2295–2301 (2020)
10. J. Webb, J.H. Holgate, *Encyclopedia of Food Science and Nutrition*, 2nd edn. (2003)
11. K.F. Mak, C. Lee, J. Hone, Jie Shan, T.F. Heinz, PRL **105**, 136805 (2010)
12. S. Xu, D. Li, P. Wu, Adv. Funct. Mater. **25**, 1127–1136 (2015)
13. D. Gopalakrishnan, D. Damien, M.M. Shaijumon, ACS Nano **8**, 5297–5303 (2014)
14. T. Wang, L. Liu, Z. Zhu, P. Papakonstantinou, J. Hu, H. Liu, M. Li, Energy Environ. Sci. **6**, 625–633 (2013)
15. B. Li, L. Jiang, X. Li, P. Ran, P. Zuo, A. Wang, L. Qu, Y. Zhao, Z. Cheng, Y. Lu, Sci. Rep. **7**, 1–12 (2017)
16. X. Ren, L. Pang, Y. Zhang, X. Ren, H. Fan, S. Liu, J. Mater. Chem. A **3**, 10693–10697 (2015)

Chapter 3
Effect of Li_2SO_4 Addition on $35Li_2O$-$65SiO_2$ Glass System

Megha A. Salorkar and V. K. Deshpande

Abstract The $35Li_2O$-$(65\text{-}x)SiO_2$-xLi_2SO_4 glass system has been synthesized by melt quenching technique in the present work. The glass transition temperature (T_g), density, XRD, Complex Impedance Measurement, and FTIR were analyzed for all prepared glass samples. This glass system shows a predominant ion-conducting nature, which was confirmed from ionic transport number measurement. The Li_2SO_4 addition in binary lithium silicate glass shows enhancement in conductivity. The T_g, density, and FTIR results have nicely supported the variation in conductivity. In this work, the 15 mol% Li_2SO_4 added glass sample shows the minimum activation energy of 0.75 eV and the maximum σ of 7.61×10^{-3} S/cm at 250 °C. It is possible to further develop this glass sample as a solid electrolyte for practical applications.

3.1 Introduction

Today, energy sources like batteries, fuel cells, and other energy storage devices are most important for portable power applications [1]. The investigation and development of safer, reliable, smaller, and higher energy storage dense batteries are in demand everywhere. Since liquid electrolytes suffers from many disadvantages such as short-circuiting, flammability, leakage, etc., many researchers have focussed their attention on solid electrolytes [2]. Among the different superionic solid electrolytes, the glassy solid electrolyte is preferable due to their stability, ionic conductivity, availability in different sizes and shapes, etc.

Manipulation of glass structure by increasing mobile ion concentration is one of the ways to increase Li-ion conductivity. The addition of lithium salts such as lithium chloride and lithium sulphate modifies the glass matrix, leading to enhancement in conductivity [3, 4]. The silica-based glasses possess excellent chemical and thermal durability but they have high melting point. This work aims to reduce the melting temperature of binary lithium silicate glass and enhance the conductivity. For this

M. A. Salorkar · V. K. Deshpande (✉)
Department of Physics, Visvesvaraya National Institute of Technology, Nagpur 440010, India
e-mail: vkdeshpande@phy.vnit.ac.in

R. G. Nair et al. (eds.), *Proceedings of 28th National Conference on Condensed Matter Physics*, Springer Proceedings in Physics 269,
https://doi.org/10.1007/978-981-16-5407-7_3

purpose, Li_2SO_4 has been added to lithium silicate glass system at the cost of SiO_2 and the change in various properties of the glass system has been studied.

3.2 Experimental Procedure

The system $35Li_2O$: $(65\text{-}x)SiO_2$: xLi_2SO_4, where $x = 0, 5, 10, 15$ mol% has been synthesized by melt quenching technique. The high graded materials Li_2CO_3 (Merck), SiO_2 (Sigma–Aldrich), and Li_2SO_4 (Loba Chemie) were used to prepare glass samples. All the materials were mixed properly, heated, and then quenched at room temperature. The quenching temperature of this glass system were in the range from 1150 to 1450 °C. All the quenched glass samples were annealed at 350 °C for two hours.

To confirm the amorphous nature of prepared glass compositions, XRD has been done by Brucker, D8 Advance Eco Powder X-Ray diffractometer. The thermal property of glass, i.e., glass transition temperature (T_g) was analyzed using differential thermal analyzer (Shimadzu, DTG-60). The density measurement for all the glass samples carried out by Archimedes principle and electrical conductivity measurement was done using Alpha Analyzer. Predominant ion-conduction in the present glass system was confirmed from DC-polarization technique and vibrations or stretching of different functional groups were analyzed by Fourier Transform Infrared Spectroscopy (Shimadzu, IRAffinity1) spectrophotometer.

3.3 Result and Discussion

3.3.1 XRD

Figure 3.1 shows the typical XRD pattern for glass sample G_3. This figure revealed the absence of characteristic peaks attributed to crystalline phases. This finding confirms the amorphous nature of the glass sample. The same type of diffractograms have been observed for remaining glass compositions.

3.3.2 Differential Thermal Analysis

The differential thermal analysis plots for all glass samples have been shown in Fig. 3.2. The first endotherm in these plots correspond to the glass transition temperature (T_g), and the first exotherm indicates the crystallization temperature (T_c). The values of T_g and T_c were tabulated in Table 3.1. The reduction in T_g with additive suggests the weakening of the glass matrix. Incorporation of SO_4 tetrahedra into the

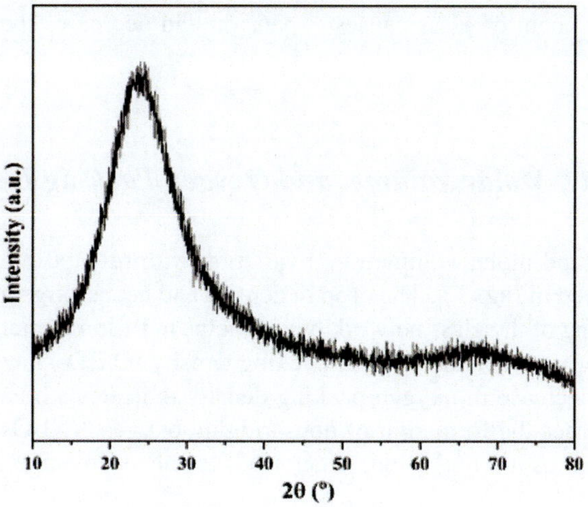

Fig. 3.1 X-Ray diffractogram for G_3 glass composition

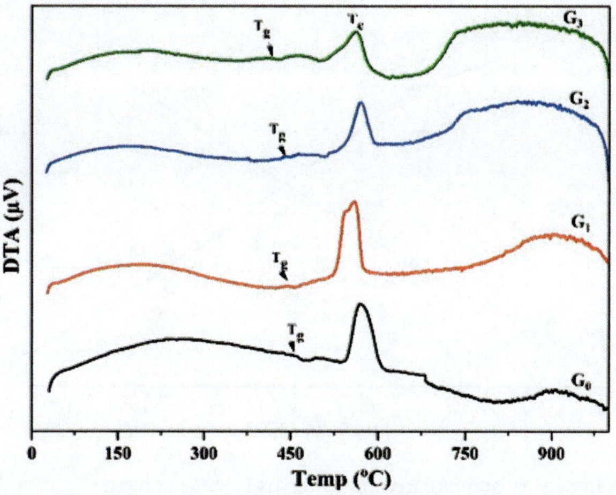

Fig. 3.2 Differential thermal analysis plot for all glass samples

Table 3.1 Values of T_g and T_c

Sr No	Li_2SO_4 content (mol %)	Sample code	T_g (°C)	T_c (°C)
1	0	G_0	449	569
2	5	G_1	442	559
3	10	G_2	433	567
4	15	G_3	418	558

macromolecular chain of glass causes expansion and decreases the rigidity of the glass matrix [5].

3.3.3 Density, Molar Volume, and Oxygen Packing Density

The density (ρ) and molar volume (V_m) variation with respect to Li_2SO_4 additive has been displayed in Fig. 3.3. The drop in density and hence surge in molar volume shows the opening of the glass network, which helps in Li-ion migration. The values of density, molar volume, and oxygen packing density (O.P.D.) have been given in Table 3.2. The decrease in oxygen packing density indicates a breaking of silicate bonds which causes the formation of non-bridging oxygens (NBOs). These NBOs provide the hopping cites for Li-ion, which leads to enhancement of Li-ion mobility.

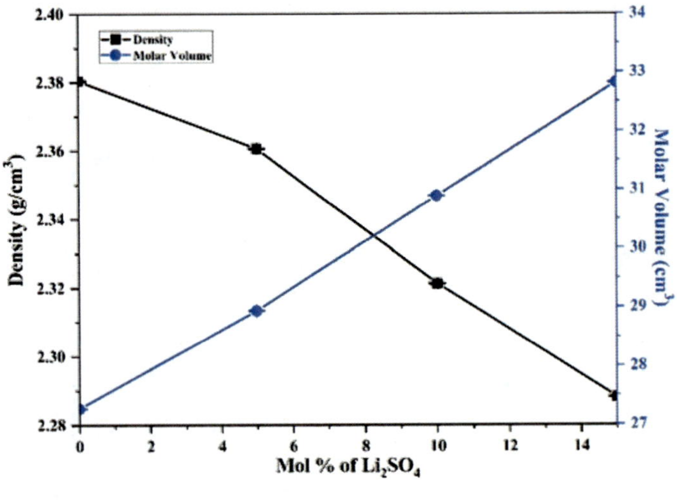

Fig. 3.3 Variation in density and molar volume versus Li_2SO_4 content

Table 3.2 Values of ρ, V_m, O.P.D. Conductivity (σ) and E_a

Sr. No	Sample code	ρ (g.cm^{-3})	V_m (cm^3)	O.P.D	σ at 250 °C (S.cm^{-1})	E_a (eV)
1	G_0	2.38	27.27	60.51	1.63×10^{-3}	0.81
2	G_1	2.36	28.94	60.48	1.98×10^{-3}	0.79
3	G_2	2.32	30.89	59.89	2.34×10^{-3}	0.77
4	G_3	2.28	32.82	59.41	7.61×10^{-3}	0.75

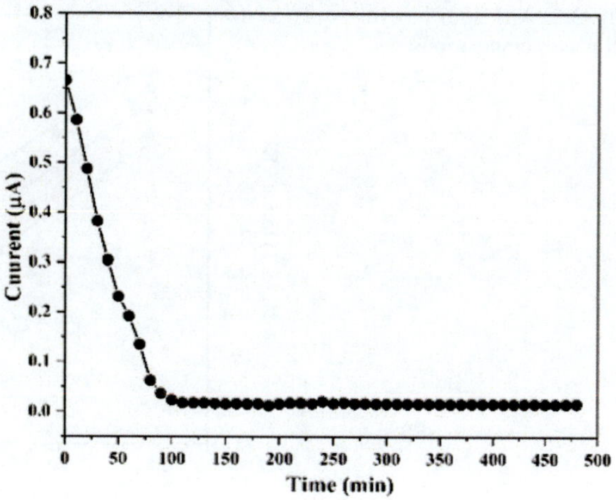

Fig. 3.4 Current versus time plot for G_3 glass composition

3.3.4 Ionic Transport Number

The DC-polarization technique involves the measurement of current at a constant supply of voltage with respect to time. The current value at zero time corresponds to total current (I_0) whereas the current that remains constant with time is known as saturation current (Is), which corresponds to electronic species. By noting these two values, ionic transport number (t_i) was calculated by the relation,

$$t_i = I_0 - I_s / I_0$$

Figure 3.4 shows the current vs time plot for glass composition G_3. Ionic transport number for the present glass system is found to be $\geq$0.992, which confirms the predominant ion-conduction nature.

3.3.5 Conductivity Measurement

The ionic conductivity (σ) of all the glass compositions was calculated in the temperature range of 100 to 275 °C from a complex impedance plot. The Arrhenius plots for all glass samples have been displayed in Fig. 3.5. This figure shows the conductivity enhancement with temperature in the measured temperature range.

The glass composition $35Li_2O$: $50SiO_2$: $15Li_2SO_4$ possesses the highest ionic conductivity in the entire series. The enhancement in conductivity has been well supported by decreased density, T_g, oxygen packing density, and increased molar volume. The values of σ at 250 °C and activation energy (E_a) for all glass samples are given in Table 3.2. This table indicates an upsurge in conductivity with lithium sulphate and hence a decrease in activation energy.

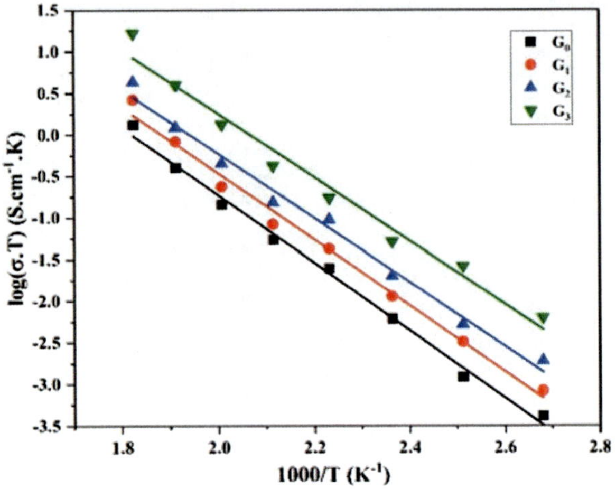

Fig. 3.5 Arrhenius plot for all glass compositions

3.3.6 Fourier Transform Infrared Spectroscopy

The Fourier transform infrared spectra for all glass compositions have been depicted in Fig. 3.6. The shoulder at about 551 and 584 cm^{-1} are assigned to O–Si–O vibrations [6, 7]. The bands at 641 to 764 cm^{-1} were attributed to symmetric stretching of Si–O–Si [6, 8]. The transmission band at 920 and 1022 cm^{-1} corresponded to νs mode of Si–O–Si and vibration of Si with one NBO [9]. These two bands become prominent with an increase in Li_2SO_4 content. This confirms the occurrence of structural variation due to additives and indicates the rise in NBO.

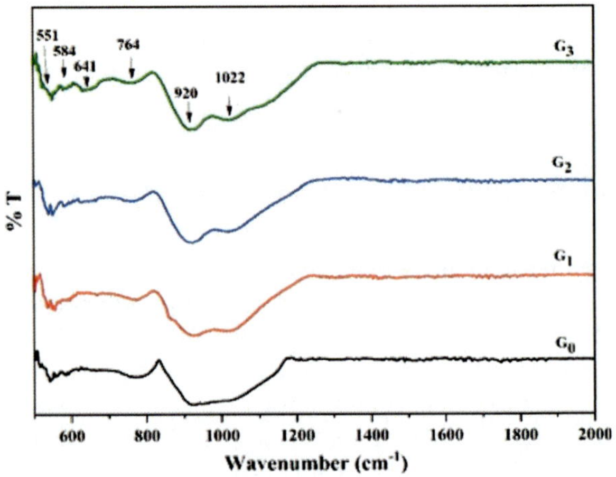

Fig. 3.6 Fourier transform infrared spectra for all glass compositions

3.4 Conclusion

The glass composition $35Li_2O$-$50SiO_2$-$15Li_2SO_4$ shows the highest ionic conductivity of 7.61×10^{-3} S.cm^{-1} among all prepared glass samples. This glass composition shows the optimum value of T_g, density, molar volume, O.P.D., and activation energy. Li_2SO_4 addition in lithium silicate glass composition causes a decrease in glass rigidity, expansion of glass matrix, increased number of vacant cites and NBOs which is well supported by DTA results, density, and molar volume, impedance measurement, and FTIR analysis.

References

1. R. Komiya, A. Hayashi, H. Morimoto, M. Tatsumisago, T. Minami solid state lithium secondary batteries using an amorphous solid electrolyte in the system (100-x)(0.6Li_2S. 0.4SiS_2).xLi_2SO_4 obtained by mechanochemical synthesis. Solid State Ionics **140**, 83 (2001)
2. Z.A. Grady, C.J. Wilkinson, C.A. Randall, J.C. Mauro, Emerging role of non-crystalline electrolytes in solid-state battery research. Front. Energy Res. **8**, 1 (2020)
3. V.C.V. Gowda, B.K. Chethana, C.N. Reddy, Ion transport studies in lithium phospho-molybdate glasses containing Cl^- ion. Mater. Sci. Eng. B **178**, 826 (2013)
4. M.A. Salorkar, K. Gour, V.K. Deshpande, Study of ion conducting $40Li_2O$-$38B_2O_3$-$20SiO_2$-$2P_2O_5$ glass system with addition of Li_2SO_4. J. Alloys Compd. **865**, (2021)
5. S.S. Gundale, V.V. Behare, A.V. Deshpande, Study of electrical conductivity of Li_2O-B_2O_3-SiO_2-Li_2SO_4 glasses and glass-ceramics. Solid State Ionics **298**, 57 (2016)
6. A. Alemi, S. Khademinia, M. Sertkol, Lithium disilicate ($Li_2Si_2O_5$): Mild condition hydrothermal synthesis, characterization and optical properties. J. Nanostruct. **4**, 407 (2014)
7. T. Fuss, A. Mogus, C.S. Ray, C.E. Lesher, R. Youngman, D.E. Day, Ex Situ XRD, TEM, IR, Raman and NMR spectroscopy of crystallization of lithium disilicate glass at high pressure. J. Non Cryst. Solids **352**, 4101 (2006)
8. A. Yang, H. Wang, W. Li, J. Shi, Synthesis of lithium metasilicate powders at low temperature via mechanical milling. J. Am. Ceram. Soc. **95**, 1818 (2012)
9. V.K. Deshpande, M.A. Salorkar, Study of lithium phosphosilicate glasses with Li_2SO_4 addition. Mater. Today Proc. **43**, 714 (2021)

Chapter 4
Effect of K_2O on the Crystallization Kinetics, Structural, Micro Structural and Mechanical Properties of Lithium Disilicate ($Li_2Si_2O_5$) Based Glasses and Glass Ceramics

Peddy Satyanarayana and A. V. Deshpande

Abstract Lithium disilicate glasses ((33.333-X) Li_2O: 66.666 SiO_2): $X K_2O$ (Where, $X = 0$, 0.5, 1 and 1.5 mol%) have been synthesized by melt quenching technique. The glasses have been converted into glass ceramics following two stage heat treatment schedules. Differential thermal analysis and Raman spectroscopy have been carried out to determine the crystallization Kinetics and structure of the glasses. The density, molar volume and microhardness measurements have been carried out for all the glasses and glass ceramics. XRD and SEM analysis have been carried out for the glass ceramics. The addition of K_2O reduces the glass transition temperature, activation energies for crystallization and percentage crystallinity. The variation in glass transition temperature, activation energy for crystallization, percentage crystallinity, density, molar volume and microhardness with the addition of K_2O has been discussed.

4.1 Introduction

Lithium disilicate glass ceramics have been grabbing the attention of researchers due to their aesthetic teeth like appearance and economic feasibility. Several authors have studied the nucleation Kinetics for stoichiometric Lithium disilicate glasses with additives such as ZrO_2, P_2O_5 [1–3]. Addition of several other oxides to Lithium disilicate based glasses has been reported to improve the mechanical properties [4]. Although the thermal and physical properties such as crystallization kinetics, density, microhardness have been studied for stoichiometric composition ($1Li_2O:2SiO_2$) of Lithium disilicate for glass ceramic manufacture [5], the glass ceramics derived from parent stoichiometric binary system possess mechanical properties which are unfavorable for several potential applications in technological areas such as dentistry and armors. Fernandes et al. [6, 7] have systematically studied the effect of K_2O on the

P. Satyanarayana · A. V. Deshpande (✉)
Visvesvaraya National Institute of Technology, Nagpur 440010, India

R. G. Nair et al. (eds.), *Proceedings of 28th National Conference on Condensed Matter Physics*, Springer Proceedings in Physics 269,
https://doi.org/10.1007/978-981-16-5407-7_4

structure and other properties such as density, coefficient of thermal expansion and bending strength of non stoichiometric Lithium disilicate glass ceramics containing Al_2O_3 by varying the SiO_2/Li_2O. The choice of this work is to study the effect of lower content of K_2O on the crystallization kinetics, density, crystallinity and microhardness of lithium disilicate glasses and glass ceramics.

4.2 Experimental

Lithium disilicate based glasses (33.333-*X*) Li_2O: 66.666SiO_2: XK_2O (With $X = 0$, 0.5, 1 and 1.5 mol%) have been synthesized by melt quenching technique. The glasses have been represented as G1, G2, G3 and G4 for $X = 0$, 0.5, 1 and 1.5 mol% K_2O respectively. The synthesized glasses and their compositions have been represented in Table 4.1. The batches taken were melted in platinum crucible at 1450 °C for 2 h and quenched into preheated copper moulds and then transferred into annealing furnace to avoid internal stresses. Differential thermal analysis (DTA) has been carried out for all the glasses to determine their glass transition temperature (T_g) and crystallization peak temperature (T_C) using simultaneous DTA-TG (SHIMADZU). The T_g and T_C measurement error was ±2 °C. DTA plots of all the glasses at heating rate of 10 K/min

Table 4.1 Nomenclature and other physical properties of glasses and glass ceramics

Glass name	G1	G2	G3	G4
$T_g \pm 2$ (°C)	465	456	453	452
$T_{C\ onset}$ (°C)	563	559	563	560
Activation energy E_C (KJ.mol^{-1})	248.17 ± 2.63	234.4 ± 2.13	214.501 ± 1.66	196.12 ± 3.36
Microhardness (H_v ± S.D) GPa	4.98 ± 0.08	4.96 ± 0.09	4.83 ± 0.11	4.76 ± 0.08
Molecular weight (W)	50.01	50.33	50.65	50.95
Density (ρ_G) ± 0.001	2.359	2.370	2.378	2.382
Molar volume V_{mG} (cm^3/mol) ± S.D	21.19 ± 0.01	21.22 ± 0.02	21.30 ± 0.01	21.39 ± 0.01
Glass ceramic name	GC1	GC2	GC3	GC4
Microhardness (H_v ± S.D) GPa	5.30 ± 0.21	5.32 ± 0.26	5.95 ± 0.27	5.29 ± 0.21
Density(ρ_{GC}) ± 0.001	2.39	2.406	2.432	2.451
Molar volume V_{mGC} (cm^3/mol) ± S.D	20.924 ± 0.01	20.918 ± 0.01	20.828 ± 0.01	20.795 ± 0.02
% Li_2SiO_3	–	2.78	6.345	9.06
% $Li_2Si_2O_5$	82.28	62.35	61.37	55.58
% Crystallinity	82.28	65.13	67.715	64.64

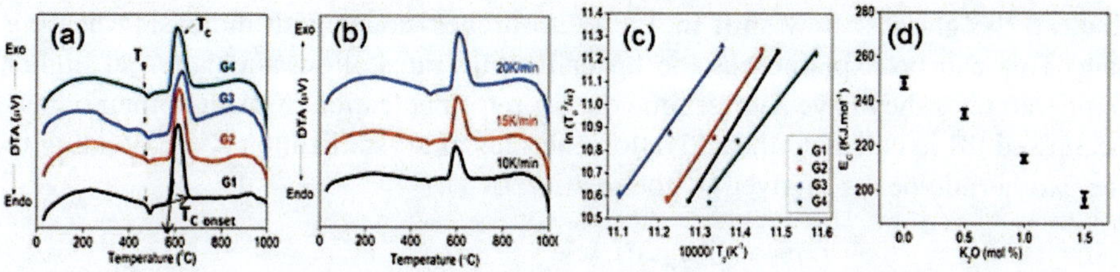

Fig. 4.1 **a** DTA plots of lithium disilicate glasses containing K_2O at a heating rate of 10 K/min and **b** DTA plots for glass G2 with different heating rates (α = 10,15 and 20 K/min); **c** Plot of $\ln(T_C{}^2/\alpha)$ versus $10{,}000/T_C$ and **d** Variation of activation energy of exothermic peak with mol% K_2O for Lithium disilicate glasses containing K_2O

have been shown in Fig. 4.1a. In this study, non isothermal method has been used to determine the activation energy for crystallization. For this purpose, all the glasses have been crushed into powder of 100–200 μm granules and then DTA has been carried out from room temperature to 1000 °C at different heating rates (α = 10, 15 and 20 $K.min^{-1}$) using platinum crucible under flowing nitrogen atmosphere. As a representative, DTA plots of glass with 1 mol% K_2O at different heating rates have been shown in Fig. 4.1b. The glasses have been cut into 4 mm × 5 mm × 10 mm ingots with the help of diamond precision saw before converting them into glass ceramics. The glasses were annealed at both the T_g and T_{Conset} (Crystallization onset temperature) temperatures respectively for 5 h. The nomenclature of glass ceramics has been represented in Table 4.1. Density (ρ) of glasses and glass ceramics has been measured by employing Archimedes's principle. From the density values, the molar volume V_m has been determined. Raman spectra were recorded for all the glasses with LABRAM HR-800 Raman spectrometer using a He–Ne laser (532 nm). The Raman data was smoothed and curve fitted. Powder XRD for all the glass ceramics has been recorded to confirm the crystalline phases present using Xpert-Pro Diffractometer equipped with a Cu K_α X-ray source in the 2θ range between 10° and 70° with a step size of 0.008°. Microstructure of the glass ceramics has been observed using FESEM (JSM-7610F) for the samples polished and then etched for 30 min in 5 volume percent Hydrofluoric acid.

4.3 Results and Discussion

4.3.1 DTA and Activation Energy of Crystallization

Figure 4.1a shows DTA plots for Lithium Disilicate glasses containing K_2O. From the DTA plots it can be observed that T_g decreases with the addition of K_2O which, indicates that there is increase in number of non bridging oxygens (NBO) [6]. The exothermic peak temperature T_C corresponds to Li_2SiO_3 phase. Figure 4.1b depicts that, all the glasses show shift in T_g, T_C to higher values with increasing heating rate. This can be explained as the faster heating rates allow nucleation at higher temperature as they have shorter time duration for nucleation. Kissinger method has been used [8] to calculate the activation energies of crystallization corresponding to the exothermic peaks as given by the relation (4.1):

$$ln\frac{T_C{}^2}{\alpha} = \frac{E_c}{RT_C} + ln\frac{E_c}{Rv} \tag{4.1}$$

T_C, α, R and ν are the peak temperature observed, heating rate used in DTA, universal gas constant and frequency factor respectively. The activation energy for crystallization 'E_C' can be determined from the slope of $ln[T_C{}^2/\alpha]$ versus $T_C{}^{-1}$. Figure 4.1c shows the plots drawn between $ln\ (T_C{}^2/\alpha)$ and $10{,}000/T_C$ and Fig. 4.1d shows the variation of activation energy (E_C) obtained corresponding to the exothermic peaks with mol% K_2O. The decrement in activation energies for all the glasses may be attributed to the weakening of the network due to the addition of K_2O [9]. Decreased activation energy E_C represents the structural role of K_2O as network modifier. The decrease in activation energy of crystallization leads to the increase of Li_2SiO_3 phase which has been discussed Sect. 3.3.

4.3.2 Density, Molar Volume and Microhardness

The variation in density and molar volume of glasses and glass ceramics with the addition of K_2O has been shown in Fig. 4.2a and b respectively. The variation in microhardness of glasses and glass ceramics with mol% K_2O has been shown in Fig. 4.2c. Molecular weight (W), density (ρ), molar volume (V_m), microhardness (H_v) and percentage crystallinity of all glasses and glass ceramics have been given in Table 4.1. The density of glasses increases with K_2O addition which can be explained on the basis of heavier K_2O (94.196 gm/mol) replacing Li_2O (29.9 gm/mol). With addition of K_2O, the molar volume increases due to the incorporation of K_2O in the interstices.

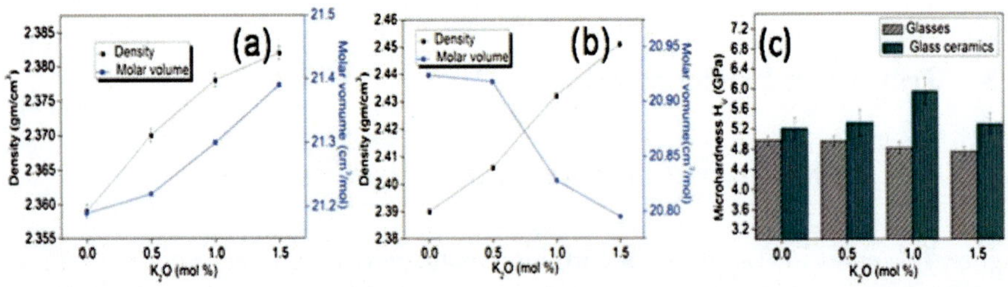

Fig. 4.2 **a,b** The variation of density and molar volume of glasses and glass ceramics with the addition of K_2O respectively. **c** The variation in microhardness of glasses and glass ceramics with the addition of K_2O

This is also supported by the observed decrease in T_g. The density of glass ceramics has been observed to be higher than their parent glasses. This can be attributed to the development and growth of crystalline phases in glass ceramics which are denser compared to the amorphous structure of the glasses. The density of glass ceramics increases with the addition of K_2O. This can be attributed to the increase in the percentage of Li_2SiO_3 phase and decrease in percentage of $Li_2Si_2O_5$ phase observed from XRD results, as Li_2SiO_3 phase is denser compared to the $Li_2Si_2O_5$ phase [10, 11]. As the density of glass ceramics increases it results in decrease in molar volume unlike glasses. Microhardness of glasses and glass ceramics has been determined using the relation; "$H_v = 1.8544(P/d^2)(kg/mm^2)$". The microhardness of glasses depends upon the rigidity of the network. Microhardness of glasses has been observed to decrease as the K_2O increased from 0 to 1.5 mol% which can be due to the weakening of the network. This is also supported by decrease in T_g. Similar results have been reported earlier [12]. Microhardness of the glass ceramics has been observed to vary in the range 5.95 ± 0.27 to 5.29 ± 0.21 GPa with the addition of K_2O. The microhardness of glass ceramics is a function of crystallinity, type of phases present and microstructure. The glass ceramic containing 1 mol% K_2O (GC3) exhibits highest microhardness 5.95 ± 0.27 GPa which may be due to larger crystal size. The decrease in microhardness of glass ceramic containing 1.5 mol% K_2O (GC4) may be due to decrease in percentage crystallinity and the formation of Li_2SiO_3 phase which exhibits lower hardness than $Li_2Si_2O_5$ phase.

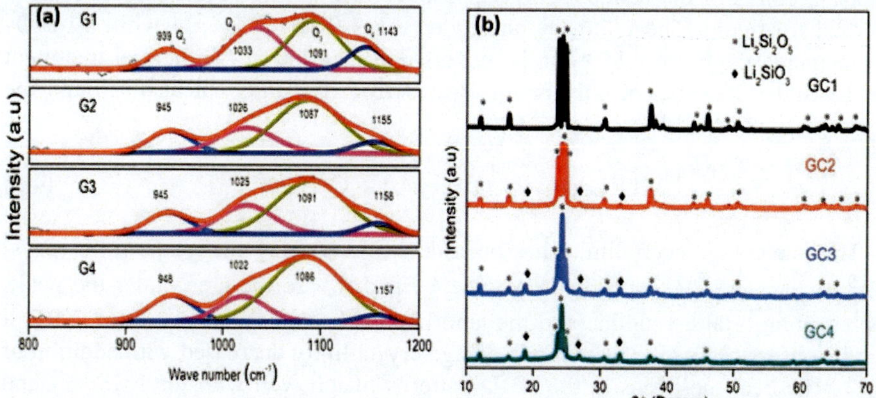

Fig. 4.3 **a** De-convoluted Raman spectra of glasses containing 0–1.5 mol% K_2O respectively, **b** XRD patterns of glass ceramics containing 0–1.5 mol% K_2O respectively

4.3.3 Structural and Micro Structural Studies

The deconvoluted raman spectra of glasses have been shown in Fig. 4.3a. The Raman spectra can be described by Q^n units (where, Q^n- Silica tetrahedra with 'n' bridging oxygens) where 'n' ranges from 0 to 4. Q^0, Q^1, Q^2, Q^3 and Q^4 represents Silica tetrahedra with 0, 1, 2, 3 and 4 bridging oxygens respectively. The peak around 939–948 cm^{-1} can be attributed to Q^2 [13]. The peak around 1022–1033 cm^{-1} can be ascribed to Q^4 units [14]. The peak around 1086–1091 cm^{-1} can be ascribed to Q^3 units and peak around 1143–1158 cm^{-1} can be ascribed to Q^4 units [13]. The glass without K_2O (G1) exhibits majorly Q^4 and Q^3 units and addition of K_2O decreases the intensity of Q^4 units while the intensity of Q^3 units increases. This indicates increase in number of non bridging oxygens. This is supported by decrease in T_g. The formation of non bridging oxygens alters the degree of crystallinity in glass ceramics which has been discussed in XRD results. XRD shows that the glass ceramic GC1 exhibits the $Li_2Si_2O_5$ phase which may be due to (1) direct formation of $Li_2Si_2O_5$ phase from Q^3 units and (2) the occurrence of Synproportionation [6] reaction (4.2) of the form;

$$Q^2\,(\text{cryst.}) + Q^4(\text{glass}) \leftrightarrow 2Q^3(\text{cryst.}) \tag{4.2}$$

where Li_2SiO_3 phase precipitates directly from Q^2 units and then reacts with amorphous Q^4 units of the residual glassy phase to form $Li_2Si_2O_5$ phase for the given thermal treatment. The $Li_2Si_2O_5$ phase decreases from GC2-GC4 while Li_2SiO_3 phase increases. From this it can be understood that during the thermal treatment the Li_2SiO_3 phase is not only being formed from Q^2 units but also through the disproportionation [6] reaction (4.3);

$$2Q^3(\text{glass}) \leftrightarrow Q^2(\text{cryst.}) + Q^4(\text{glass}) \tag{4.3}$$

The percentage crystallinity has been determined using the relation; "% Crystallinity $= \frac{\sum_{ci} A_{ci}}{\sum_{ci} A_{ci} + A_a} \times 100\%$". Where, A_{ci} and A_a are the area under the peaks representing total crystalline and the amorphous phase respectively. From XRD it can be observed that the overall percentage crystallinity decreased with addition of K_2O. The peak positions in the XRD patterns match well with the $Li_2Si_2O_5$ and Li_2SiO_3 crystals (ICSD card No.72-0102 with space group C1c1 and 83-1517 with space group $Cmc2_1$ respectively). Figure 4.3b shows that, the samples without K_2O (GC1) show $Li_2Si_2O_5$ as main phase. As the K_2O content increased the intensity of $Li_2Si_2O_5$ peak decreased and the intensity of Li_2SiO_3 peak increased. Figure 4.4a–d show the SEM images of glass ceramics with the addition of K_2O. The glass ceramic without K_2O (GC1) exhibits large number of rod like smaller crystals and the addition of K_2O changes the shape of the crystals into ellipsoids. The formation of the $Li_2Si_2O_5$ phase from both the Q^3 units and via synproportionation reaction simultaneously may result in more number of crystals in GC1. The decrease in packing of

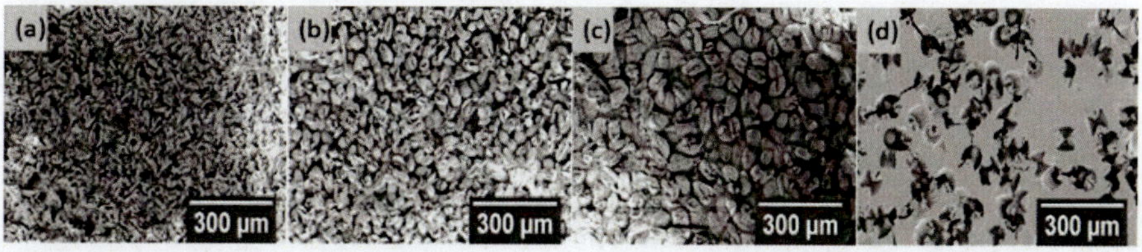

Fig. 4.4 SEM images of glass ceramics containing **a** 0 mol% K_2O, **b** 0.5 mol% K_2O, **c** 1 mol% K_2O and **d** 1.5 mol% K_2O respectively

crystals can be explained by the absence of simultaneous formation of Li_2SiO_3 and $Li_2Si_2O_5$ nuclei. The small number of nuclei results into growth of larger size crystals as observed from Fig. 4.4a–d. SEM image of GC4 shows etched pits corresponding to Li_2SiO_3 crystals as Li_2SiO_3 crystals are easily soluble in HF.

4.4 Conclusions

The glass transition temperature and activation energies of crystallization of Lithium disilicate glasses decrease with addition of K_2O. This confirms the role of K_2O as network modifier. The addition of K_2O increases Li_2SiO_3 phase while $Li_2Si_2O_5$ phase decreases. With increasing (K_2O/Li_2O) the glass network weakens. Addition of K_2O to Lithium disilicate system decreases the microhardness of glasses while glass ceramic containing 1 mol% K_2O exhibits highest microhardness 5.95 ± 0.27 GPa which has potential for dental and armor applications.

Acknowledgements The authors wish to thank Prof. V. K. Deshpande for providing laboratory facilities and DRDO for providing grant to support this research.

Funding This work was supported by the DRDO (Grant no. ARMREB/MAA/2016/187).

References

1. K. Thieme, C. Ru¨ssel, Nucleation and growth kinetics and phase analysis in zirconia-containing lithium disilicate glass. 1488–1499 (2015). https://doi.org/10.1007/s10853-014-8710-5
2. S. Huang, Z. Huang, W. Gao, P. Cao, Trace phase formation, crystallization kinetics and crystallographic evolution of a lithium disilicate glass probed by synchrotron XRD technique. 2–6 (2015). https://doi.org/10.1038/srep09159
3. B. Galvão, M. Valério, P.A. Suzuki, et al., Journal of the Mechanical Behavior of Biomedical Materials Mechanical properties of lithium metasilicate after short-term thermal treatments. J. Mech. Behav. Biomed. Mater. **98**,179–186 (2019). https://doi.org/10.1016/j.jmbbm.2019.06.011

4. T. Moreira, B. Campos, R. Marques et al., Impact of crystallization firing process on the microstructure and flexural strength of zirconia-reinforced lithium silicate glass-ceramics. Dent Mater **34**, 1483–1491 (2018). https://doi.org/10.1016/j.dental.2018.06.010
5. W. Höland, G. Beall, Glass ceramic technology. Am. Ceram. Soc. (2002)
6. H.R. Fernandes, D.U. Tulyaganov, A. Goel, José, M.F. Ferreira, Effect of K_2O on structure – property relationships and phase transformations in Li_2O–SiO_2 glasses. J. Eur. Ceram. Soc. **32**, 291–298 (2012). https://doi.org/10.1016/j.jeurceramsoc.2011.09.017
7. H.R. Fernandes, D.U. Tulyaganov, A. Goel, et al., Effect of Al_2O_3 and K_2O content on structure, properties and devitrification of glasses in the Li_2O–SiO_2 system. J. Eur. Ceram. Soc. **30**, 2017–2030 (2017). https://doi.org/10.1016/j.jeurceramsoc.2010.04.017
8. Y. Shaharyar, J.Y. Cheng, E. Han, et al., Elucidating the effect of iron speciation (Fe_2+/Fe_3+) on crystallization kinetics of sodium aluminosilicate glasses. J. Am. Ceram. Soc. **99**(7), 2306–2315 (2016). https://doi.org/10.1111/jace.14239
9. X.Z. Guo, H. Yang, M. Cao, C. Han, F.F. Song, Crystallinity and crystallization mechanism of lithium aluminosilicate glass by X-ray diffractometry. Trans. Nonferrous Met. Soc. China **16,** 593–597 (2006). https://doi.org/10.1016/S1003-6326(06)60104-0
10. H. Von, SEEMAN (1956) Die Kristallstruktur des Lithiummetasilikates, (Li2SiO3)x. Acta Cryst. **9**, 251 (1956). https://doi.org/10.1107/S0365110X56000693
11. B.H.W.S. De Jong, H.T.J. Supèr, A.L. Spek, N. Veldman, G. Nachtegaal, J.C. Fischer, Mixed alkali systems: structure and ^{29}Si MASNMR of $Li_2Si_2O_5$ and $K_2Si_2O_5$. Acta Cryst. B **54**, 568–577 (1998). https://doi.org/10.1107/S0108768198001062
12. S. Sen, A. Ghosh, Structural properties of strontium vanadate glasses. J. Mater. Res. **15**, 995–999 (2000). https://doi.org/10.1557/JMR.2000.0142
13. B.G. Parkinson, D. Holland, M.E. Smith, C. Larson, J. Doerr, M. Affatigato, S.A. Feller, A.P. Howes, C.R. Scales, Quantitative measurement of Q 3 species in silicate and borosilicate glasses using Raman spectroscopy. **354**, 1936–1942 (2008). https://doi.org/10.1016/j.jnoncrysol.2007.06.105
14. C. Calahoo, J.W. Zwanziger, I.S. Butler, Mechanical-structural investigation of ion-exchanged lithium silicate glass using micro-raman spectroscopy. J. Phys. Chem. C **120**, 13, 7213–7232 (2016). https://doi.org/10.1021/acs.jpcc.6b01720

Chapter 5
Enhancement of Electrical Conductivity in Nanostructured Metal Oxide Composite

Meenakshi Srivastava, Piyush Jaiswal, and Narendra Singh

Abstract Nanomaterials have been recognized for their excellent thermal, optical, electrical properties with robust machine-driven strength. The oxide and nanocomposite have been prepared using n-type semiconductor material i.e., tin oxide and zinc oxide through Sol–Gel technique. The materials were further characterized using X-ray diffraction (XRD), Scanning electron microscope (SEM), Brunauer–Emmett–Teller (BET), Fourier Transform Infrared Spectroscopy (FT-IR), etc. The XRD analysis of metal oxide and nanocomposite have been performed to determine the crystallite size and structural phase and it confirms the presence of tetragonal phase for tin oxide and hexagonal phase for zinc oxide. At the same time, composite material contains both the phases. The different morphology was observed for the tin oxide, zinc oxide and composites materials using the SEM images. With the help of SEM characterization, the surface morphology of oxide and nanocomposites have been investigated individually where the agglomeration of particles as nanostructure for tin oxide, nanorod formation for zinc oxide, and flakes like structure for oxide composites etc., have been also observed. The BET analysis was carried out for the tin oxide, zinc oxide and composite materials to find out the specific surface area and pore size distribution. It was found that composite of tin oxide and zinc oxide showed the highest BET surface area evaluated as 47 m^2g^{-1} in comparison to pure ZnO nanoparticle and SnO_2 nanomaterial. The conductivity for metal oxide and nanocomposite has been measured to predict the sensing performance of the materials.

M. Srivastava (✉) · P. Jaiswal · N. Singh (✉)
Department of Nanoscience and Technology, Centre for Advanced Studies, Dr. A.P.J. Abdul Kalam Technical University, Lucknow 226031, India
e-mail: narendra@cas.res.in

N. Singh
Department of Chemical Engineering, Indian Institute of Technology Tirupati, Tirupati, Andhra Pradesh 517506, India

R. G. Nair et al. (eds.), *Proceedings of 28th National Conference on Condensed Matter Physics*, Springer Proceedings in Physics 269,
https://doi.org/10.1007/978-981-16-5407-7_5

5.1 Introduction

Over the last few decades, environmental pollution has declined the growth and development of industrialization and urbanization. Some toxic gases such as NO_x, CO, phosgene have affected the human health and environment even at ppm levels. The nanoscale devices and technology have a swift progress and experiencing impact on each subject area of research in the decade of twenty-first century. The scientific research has shown that the exposure of toxic gases like NO_x has a harmful effect on respiratory organ such as inflammation where health is exposed and it is co-related to presence of excess of fluids in the lungs and may cause death [1–3]. The solid-state based sensing devices have been enacted in examining the chemical procedure, environmental affects, health, and safety along with medical devices. SnO_2 and ZnO being n-type semiconductor metal oxide with band gap of 3.67 eV and 3.37 eV was found to be the most prominent material for featuring device responses.

Compared with other solid state based gas sensor, metal oxide semiconductor (MOS) based gas sensor such as ZnO, SnO_2, TiO_2, and WO_3 etc., have proven the best metal oxides due to their domestic and industrial applications in poisonous and flammable gas sensing. Among the variety of MOS based sensors, SnO_2 have proved to be tremendous semiconducting sensitive material owing to its promising features likely to have large surface area, higher sensitivity, quick response time with reliable stability and in the detection of explosive and harmful gases [4].

With improvement in the potential of SnO_2 based nanostructured materials [5], ZnO based nanoparticles [6], and for nano-composites [7], different kinds of morphologies of metal oxides and composites have been observed such as nanowires [8, 9], nano-belts [10], flower like structure [5, 11], nano-pellets [12], snowflakes [12, 13] etc. From the perspective of sensor response, the factors such as selectivity, response time, sensitivity, stability and durability are the most important properties for selecting gas sensors. For detecting the target gases at low concentration, sensitivity and response of the sensor must be high which is crucial in sensing applications. The sensors based on semiconductor have been shaped from the behaviour of adsorption and desorption of gas molecules, surface sites and active regions. The nanostructured materials found to have high density which is a standard feature for sensing applications [14–16]. The hydrolysis, precipitation and heat treatment are the prominent factors for identifying the final outcome in the sol–gel procedure. Sol–Gel process has several advantages in comparison to other techniques for synthesizing the metal oxide based nanostructured materials [17]. In the present work, sol–gel routing has been chosen for the preparation of samples. The precursors as tin oxide dihydrate, zinc chloride, ammonia etc., and CTAB as surfactant with ethanol as solvents have been used. It is known that the gas sensing features have been influenced more towards grain size, morphology and active sites. The conductivity performance have been investigated for metal oxide and its composite to predict the gas sensing applications, where the powdered sample have been prepared in the pellets form to identify the electrical conductivity property with the help of electrode contacts [18, 19]. On the basis of electrical study of the metal oxide semiconductors (MOS), the detection or sensor functioning can be estimated and it is found that the MOS have the better stability and other features which could be advantageous in the improvement of sensing applications [20].

5.2 Materials and Methodology

5.2.1 Preparation of Pure SnO_2

SnO_2 based nanomaterial have been prepared using simple sol–gel method. SnO_2 was prepared using 1.21 g $SnCl_2.2H_2O$ as precursor and it was diluted in the 50 mL of ethanol and magnetic stirrer have been employed to stir the mixture for about 10–20 min while the pH was maintained at 8 by adding ammonia dropwise. A white precipitate were collected by filtering over filter paper and washed thoroughly with ethanol. The precipitate was heated at 80 °C for 4 h in a vacuum oven and further heat treatment were carried out at 450 °C for 2 h in a muffle furnace. A white powder was collected after post heat-treatment process.

5.2.2 Preparation of Pure ZnO

ZnO based nanoparticle has also been synthesized using sol–gel routing technique. ZnO is prepared using 2.20 g of dry purified form of $ZnCl_2$ with help of 50 mL of ethanol. It was kept for stirring using magnetic stirrer for about 10–20 min and ammonia was added drop by drop for maintaining pH to 8 (approximate value). Then, the precipitate was filtered and washed using ethanol and kept for pre-heat treatment in hot air oven at 80 °C for 4 h and for post heat treatment, it was kept at 450 °C for 2 h.

5.2.3 Preparation of ZnO-SnO_2(1:2) Nanocomposite

ZnO/SnO_2 based nanocomposites have been prepared with different weight ratio (1:2) using dry purified form of 1 g $ZnCl_2$ with 2 g $SnCl_2.2H_2O$ and 0.4 g of cationic surfactant CTAB (cetyltrimethyl ammonium bromide) and dissolved in 50 mL of ethanol. Further, it has been kept for continuous stirring on a magnetic stirrer for about 10–20 min and pH has been maintained to 11 for 1:2 weight ratios by adding ammonia dropwise. The precipitate has been filtered thoroughly with filter paper and washed with ethanol. Then, the sample was retained for pre-heating at 80 °C for 4 h and for post heating at 450 °C for 2 h respectively.

5.3 Characterization and Conductivity Measurement

The morphological study of the solid products was analysed using scanning electron microscopy (SEM-JEOL 7610F) along with coating lid {JEOL-JEC(3000FC)}. The energy dispersive spectroscopy (EDS) was accomplished associated with scanning electron microscope (SEM-JEOL 7610F). The EDS analysis for the elemental composition of the material was recorded using EDS (133) with EV Dry Detector (INCA x-act) of OXFORD instruments. Crystal structure was analysed by X-ray diffraction (XRD- Diffrac.Eva evaluation software, Diffrac.Topas software and ICDD PDF-4 Axiom 2020 database) using CuKα radiation, wavelength 0.15708 nm. The sample was scanned from 20° to 80° (2θ) in steps of 0.02°. The calculation of specific surface area was performed by Brunauer–Emmett–Teller (BET) method through nitrogen adsorption–desorption using BELSORP MINI X instrument. The FTIR spectrum in the range of 4000–400 cm^{-1} was performed on Perkin–Elmer Spectrum GX infrared spectrophotometer.

The conductivity measurements have been done by preparing the powdered sample in the form of pellets where contacts have been made for examining the electrical properties of the metal oxide and composites. The measurements are accomplished in the air ambiance at room temperature using a Keithley 6517B electrometer, which can measure resistance upto 10^{15} Ohms. Gold coated tungsten probes of 10 μm tip size were used to make contact with the sample in Linkam P600 stage. The gold electrode material has been used for the conductivity measurements of nanomaterials.

5.4 Results and Discussion

5.5 XRD Analysis

The XRD pattern of synthesized pure SnO_2 nanomaterial has shown the characteristic peaks of SnO_2 with tetragonal phase (JCPDS no. 41–1445) as shown in Fig. 5.1a. The lattice constants estimated for the tetragonal phase for SnO_2 has been found equal to a = b = 0.47 nm and c = 0.33 nm using the equation $1/d^2 = h^2/a^2 + k^2/b^2 + l^2/c^2$ where a, b and c are the unit cell specifications and d is the interplanar distance [21]. No other crystalline phases/impurities have been found in the synthesized materials. The intense peaks have clearly shown the formation of the tetragonal phase for pure SnO_2 having a major (1 1 0) plane in the following sample, which was found to be more stable for SnO_2 [22]. The crystallite size calculated using Debye Scherer Equation (5.1) has been found between 8 nm and10 nm.

$$D = \frac{0.9\lambda}{\beta \cos\theta} \tag{5.1}$$

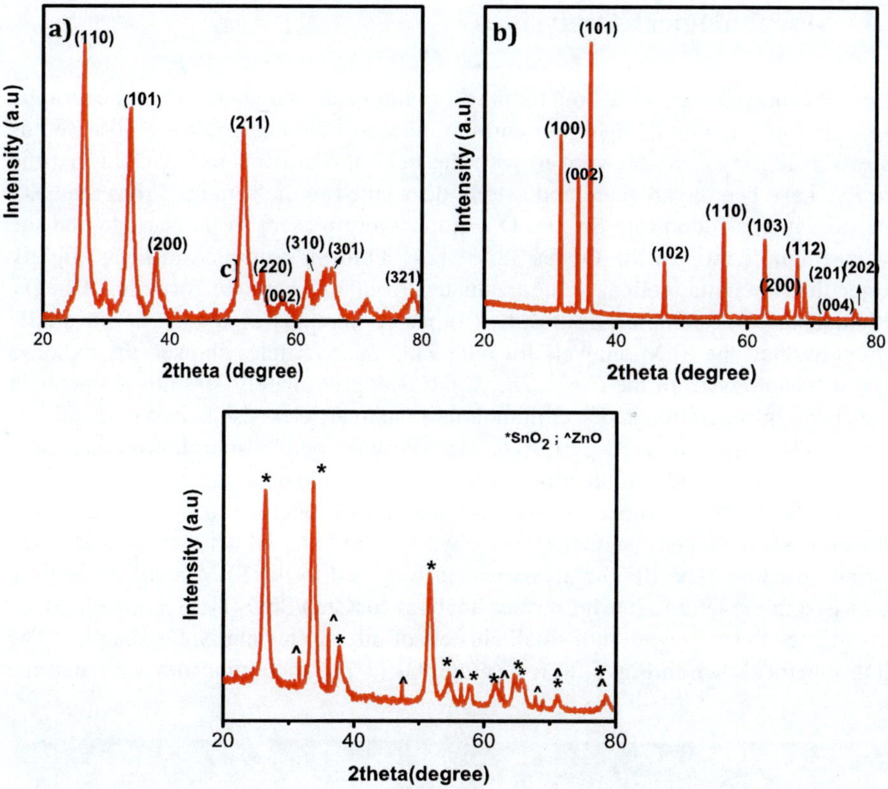

Fig. 5.1 XRD pattern of **a** pure SnO_2 nanomaterial **b** pure ZnO nanoparticle and **c** ZnO/SnO_2 (1:2) nanocomposite

D refers to crystallite size, β is full width at half the maximum of the peaks, θ known for Bragg's angle, λ is the X-ray wavelength. The prepared pure ZnO sample's structural identification was studied using XRD graph from Fig. 5.1b. The diffraction pattern showed the hexagonal phase of ZnO and its peaks were matched with JCPDS no. 01–079–208. The XRD peaks were found narrow with high intensity, which is a sign of good crystallinity. The crystallite size calculated from the Debye Scherer equation for pure ZnO nanoparticles in the range of 30–50 nm. Similarly, the phase identification and structural properties of ZnO/SnO_2 (1:2) nanocomposite studied using XRD pattern illustrated in Fig. 5.1c. The diffraction pattern for the ZnO/SnO_2 (1:2) nanocomposite consists of both ZnO and SnO_2 material structure features. The peaks are well- matched with rutile tetragonal and wurtzite hexagonal phases (JCPDS no. 41–1445 and JCPDS no. 01–079–208) [23]. The intensity of ZnO diffraction peaks was weaker than SnO_2 as the weight ratio of ZnO/SnO_2 is 1:2. No other additional peaks have been observed for confirming the existence of Zn/Sn nanocomposite with more purity. The atomic ions of Zn^{2+} and Sn^{4+} have been found as 0.074 nm and 0.069 nm.

5.6 Morphological Analysis

The SEM analysis was carried out for the morphological characteristics of the sample. For pure SnO_2 materials, it was found that nanoparticle clusters were observed as shown in Fig. 5.2a. It was viewed with the help of SEM image Fig. 5.2a that the particle have been aggregated and evenly distributed in the sample. From the EDS analysis, it was found that Sn and O's atoms were present in the sample and the average atomic ratio of Sn: O was 32: 68 [24]. O to Sn's atomic ratio was slightly larger than their theoretical value evaluated from its chemical formula of SnO_2, although there is complete removal of other compounds in the form of volatile gases. Further, the SEM analysis for pure ZnO nanoparticles showed the nanorod type morphology from the Fig. 5.2b, which is approximately 100 nm diameter in size. The agglomeration and accumulation of nanoparticles results into high surface energy. The nanostructured pure ZnO material with porous morphology has been reliable for gas sensing applications [25]. The morphological structure showed that the ZnO/SnO_2 (1:2) nanocomposite samples having aggregation of particles and formed the aggregated nanoparticles shown in Fig. 5.2c, which also indicated the porous structure. The BET analysis was also agreed to the SEM analysis, will be discussed in the later part. The surface analysis for ZnO/SnO_2 (1:2) nanocomposite investigated the aggregation of small clusters of mixed ZnO and SnO_2 samples. The EDS spectra shown in Fig. 5.2d for ZnO/SnO_2 (1:2) nanocomposites validated the

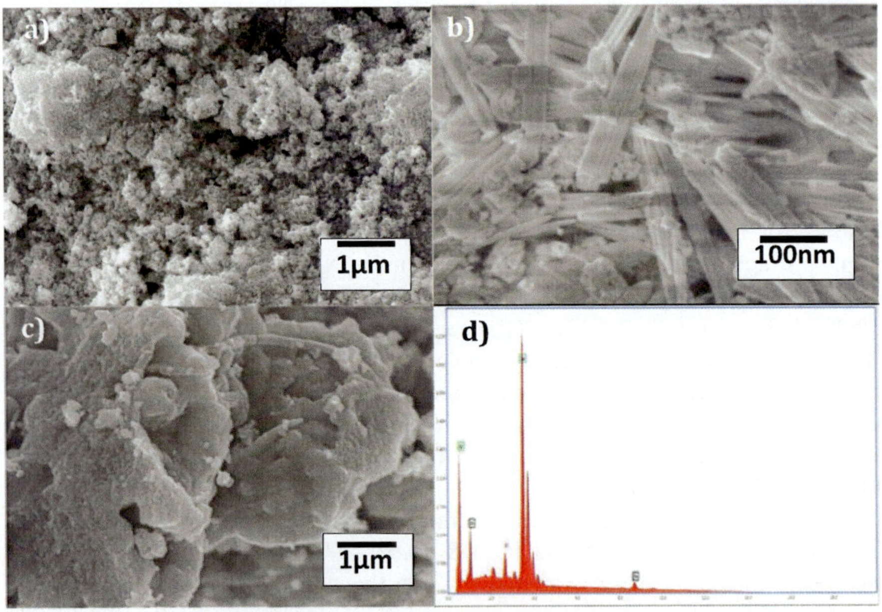

Fig. 5.2 SEM images of **a** pure SnO_2 nanomaterial **b** pure ZnO nanoparticles, **c** ZnO/SnO_2 (1:2) nanocomposite and **d** EDS analysis of ZnO/SnO_2 (1:2) nanocomposite

Table 5.1 EDS analysis for Zn/Sn (1:2) nanocomposite

Element	Weight%	Atomic%
Sn L	65.76	24
Zn K	5.39	03
O K	26.34	70
Cl K	2.51	03

presence of Zn, Sn, Cl and O elements with some impurities present in it as shown in the Table 5.1.

5.7 FT-IR Analysis

The FT-IR spectra for the prepared pure SnO_2 material has been analysed after annealing the sample to 450 °C. For the pure SnO_2 sample, the peaks were found at 622 cm^{-1} and 560 cm^{-1} due to anti- symmetric Sn–O–Sn stretching vibrations on the oxide associating surface formed by the concentration of adjacent hydroxyl group as depicted in Fig. 5.3a. The hydroxyl group attributed the absorption bands at 1622 cm^{-1} and it may be due to the re-adsorption of water molecules in atmospheric conditions [26]. Thus, these absorption bands confirmed the presence of SnO_2. The FT-IR spectrum for pure ZnO was observed from the spectra range of 4000–500 cm^{-1} to check the functional groups accompanied with ZnO. The small peaks located at 468 and 915 cm^{-1} showed the stretching vibrational mode of Zn–O bond. The peaks at 1428 and 1648 cm^{-1} were found to be due to the O–H bond and Zn–Cl stretching vibrations depicted in Fig. 5.3b. The peaks located at 2923 and 3440 cm^{-1} matched the asymmetric C–H stretching and stretching vibrations of the O–H bond in the ZnO sample. The absorption band revealed the wurtzite ZnO phase when calcined to 450 °C. Figure 5.3c shows the result of the FT-IR spectrum for prepared ZnO/SnO_2(1:2) nanocomposite. The FT-IR spectrums have been investigated in the spectra range of 4000–500 cm^{-1}. The spectrum analysis showed the absorption peaks observed at 604 and 507 cm^{-1}, which belongs to the Zn–O–Sn bond formation in the ZnO–SnO_2 (1:2) nanocomposite. The broad absorption peak at 3487 and 1620 cm^{-1} showed the band stretching of O–H and bending vibration of water molecules. As the composite annealed at 450 °C for 2 h, it was found that the bands attributed to CTAB surfactant disappeared or there have been entire elimination of surfactant from the mesostructured nanocomposite during the heat treatment [27]. From the peaks, it was confirmed that oxide has been present for synthesized ZnO–SnO_2 (1:2) nanocomposite.

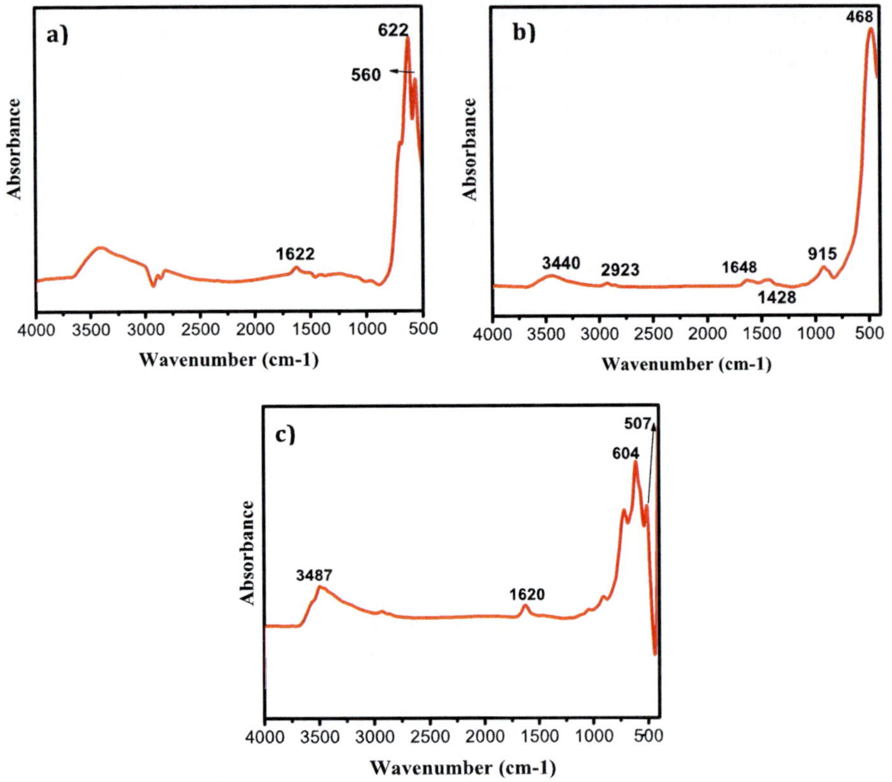

Fig. 5.3 FTIR spectra of **a** pure SnO_2 nanomaterial, **b** pure ZnO nanoparticle and **c** ZnO/SnO_2(1:2) nanocomposite

5.8 BET Analysis

The texture properties and porosity of pure SnO_2 nanomaterial have been analysed with the help of nitrogen adsorption–desorption measurement system. The adsorption–desorption isotherm curve of pure SnO_2 has analysed the hysteresis loop of type IV isotherm where adsorption for mesoporous entities following capillary condensation relates to filled mesopores and empty spaces and confirmed that it has high porosity and having narrow size distribution of pores [14]. The small intense peak of pore size was obtained in the range of 1–4 nm from the BJH plot Fig. 5.4a. The BET surface area of pure SnO_2 material was found to be 21 $m^2g^{-1}$with porous structure. From the Fig. 5.4b, the result of BJH plot for pure ZnO nanostructure showed that the micropores have been filled with gas at low pressure, and it was analysed that the pore size distribution with two-centered peaks located at 2.6 nm and 3.4 nm with an average size of 3 nm. The pore volume was low due to the calcination process, which breaks the pores and aggregation of particles and reduces

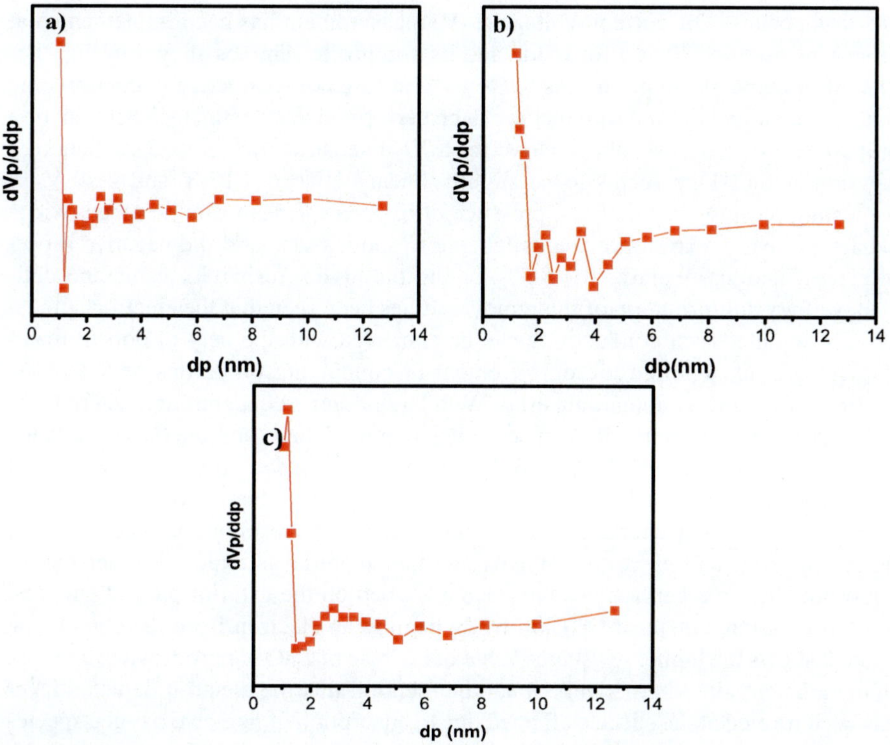

Fig. 5.4 BJH plot of **a** pure SnO_2 nanomaterial, **b** pure ZnO nanoparticle and **c** ZnO/SnO_2 (1:2) nanocomposite

the specific surface area. Further, the plot of BJH shown in Fig. 5.4c and adsorption–desorption isotherm measurement for porous ZnO/SnO_2 nanocomposites with different ratio (1:2), it has been studied that hysteresis loop of type IV shows the multilayer adsorption subsequently capillary condensation associated with presence of filled and empty mesopores with diameter of 2–50 nm approximately. The surface area for Zn/Sn (1:2) composite was found to be 47 m^2g^{-1} which confirms the higher porosity than the pure samples [23].

5.9 Electrical Conductivity Measurement

The electrical measurements were carried out to investigate the metal oxide and composites' electrical conductance changes based on the current–voltage curve. It is known that metal oxide semiconductor is responsive to oxygen in the surroundings and variation in the conducting behaviour can be measured. The sensing can be analysed with the help of electrical characterization measurements for nanomaterials and nanocomposites. The current–voltage (I–V) measurement has been implemented at room temperature (30 °C) for oxide and its composite samples shown in Fig. 5.5. The metal oxide of n-type such as SnO_2 and ZnO are not conductive in their pristine states, so conductivity measurements can be performed to investigate the conducting properties. The current–voltage curve for SnO_2 nanostructure showed the Schottky junction as the slope seems to be slightly linear between +10 V and +20 V. In the testing of SnO_2 material, the presence of oxygen species from the metal oxides behave as charge carriers for electrochemical applications and the negative aspect of electrochemical features for SnO_2 having the higher resistivity value and deficiency of crystal formation of ion removal. It has been seen that the electrochemical activity and electrical conductivity can be enhanced with the help of dopant material. Jiang et al. reported the improvement of conducting properties of SnO_2 fine coatings with zinc as dopant material. With large doping concentration, Zn reduces the carrier concentration with respect to behaviour of SnO_2 and the band gap value found to be low from 3.93 eV to 3.79 eV caused from the drop in value of ZnO in comparison to SnO_2 [28]. For pure ZnO nanoparticle the curve illustrated the linear non-rectifying characteristics which form the good ohmic contact between device material and electrode. When ZnO nanomaterial is exposed under the air atmosphere, an oxygen species has been adsorbed on the exterior part of ZnO and there is an arrangement of $O^{-}{}_2$ ion while gaining an electron from the conduction band. It shows the higher resistance value in the case of ZnO nanorods when exposed in the ambient air. Alternatively, it can be mention that the electron depleted layer has been formed at the surface affected due to adsorption of negative oxygen species at an ambient temperature where the electrons have been obtained in the upper band, increasing the barrier region and resistance value of the sensor. It was seen that ZnO's slope is sharper compared to pure SnO_2 nanostructure as the forbidden gap of ZnO is narrower than SnO_2. Similar results for conductivity performance for nanomaterials can be observed by various researchers. H. Xu et al. investigated the enhancement in the performance of ZnO gas sensors where it has been found that the crystallinity

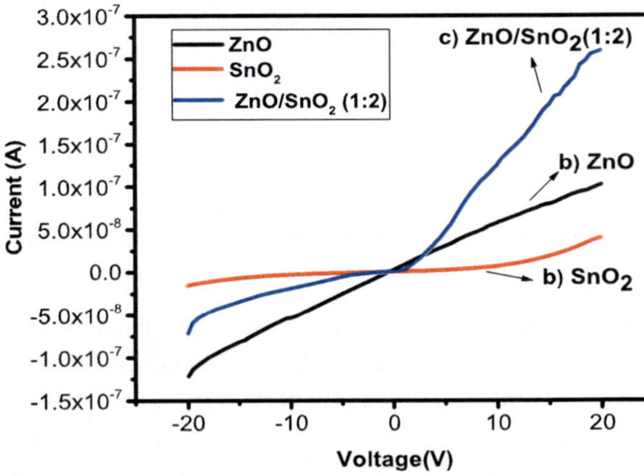

Fig. 5.5 Current–voltage measurement of **a** pure ZnO nanomaterial, **b** pure SnO_2 nanomaterial and **c** ZnO/SnO_2 (1:2) nanocomposite

of ZnO has been improved due to its porous nature with high intrinsic resistance value and poor selectivity feature [29]. The I-V characteristics of ZnO/SnO_2 (1:2) nanocomposite showed the linear behaviour as in the Fig. 5.5. The conductivity indicated that from 0 V onwards, the device can operate easily with a good signal to noise ratio. In general, the I-V curve shows nonlinearity for metal-oxide composite considerably and also the heterojunction of n–n type, thus formed having a distinct height of built-in potential between electron and hole. The change in the grain boundaries built-in potential of the composites, have been instigated by the formation of heterojunctions. In the case of ZnO–SnO_2, there is a formation of n–n heterojunction of the material which acts as an key part in the improvement of the sensor performance. The following paper reported the thin film based ZnO-SnO_2 nanocomposite sensor in the detection of H_2 gas, where the sensing response was found to be high i.e., 8.8×10^2 at the lower operating temperature of 160 °C. It was also found that there is a declining in the resistance value towards the target gas [30]. B. Mondal et al. investigated the detection of hydrogen gas with the help of chemical route of ZnO–SnO_2 nanocomposite where 90% response of the sensor was observed in the 1000 ppm concentration of hydrogen gas at150 °C. It indicated that the sensor has a higher value of reluctance in the air ambience and even more resistivity value when exposed to gases [31].

5.10 Conclusions

Sol-Gel technique has been used for synthesizing the metal oxide and its nanocomposite using CTAB as surfactant. This bottom up approach is simple, convenient to use, and environment friendly and efficient in preparation of nanoscale material. A suitable amount of precursors, CTAB as surfactant, additives and ethanol as solvent have been used at an initial phase for the preparation of nanostructured material in different weight ratios as required. The structural analysis was performed using XRD which confirmed the occurrence of tetragonal phase for tin oxide and hexagonal phase for zinc oxide and mixture of both phases for nanocomposites. The microstructural analysis has been observed with the help of SEM which validates the nanostructured morphology which was found to be suitable for sensing application. The spectroscopic analysis for metal oxide and its composite have investigated about the Sn–O–Sn stretching vibrations, Zn–Cl bonding, N-CH_3 symmetric and asymmetric C–CH_3 stretching vibrations of solid surfactants, Zn–O–Sn bonding, and O–H bending of hydroxyl group. The surface area measurement has been examined using BET analyser which predicted that the metal oxide nanocomposite (1:2) has the highest BET surface area among other oxides. The electrical properties for metal oxide and its composite have been observed through conductivity measurement which revealed that the I–V curve for ZnO/SnO_2 nanocomposite showed the linearity with good ohmic contact and have better conductivity features which can enhance the sensor response for further device applications. From the perspective of surface area and porosity, the nanocomposite found to have high specific surface area and good textural porosity which is prominent for sensing applications.

Acknowledgements The authors are thankfully appreciated the support granted from the Department of Nanoscience & Technology, Centre for Advanced Studies, Dr. APJ Abdul Kalam Technical University, Lucknow under Uttar Pradesh State Government and would like to thanks and acknowledge the support of Dr.Vinayak Kamble for the conductivity testing facility under CIF instrumentation lab facility of IISER, Thiruvananthapuram.

References

1. N. Barsan, D. Koziej, U. Weimar, Metal oxide-based gas sensor research: How to? Sens. Actuators B: Chem. **121**(1), 18–35 (2007)
2. M. Hasan et al., A review on the pattern of electricity generation and emission in Indonesia from 1987 to 2009. Renew. Sustain. Energy Rev. **16**(5), 3206–3219 (2012)
3. X. Ji, G. Chen, Unified account of gas pollutants and greenhouse gas emissions: Chinese transportation 1978–2004. Commun. Nonlinear Sci. Numer. Simul. **15**(9), 2710–2722 (2010)
4. N. Barsan et al., Chemical and biochemical sensors, 1. Fundamentals. Ullmann's Encyclopedia of Industrial Chemistry. Wiley-VCH,1–81, (2016)
5. J. Hu et al., Hierarchical aloe-like SnO nanoflowers and their gas sensing properties. J. Mater. Res. **33**(10), 1433–1441 (2018)
6. K. Ghule et al., Preparation and characterization of ZnO nanoparticles coated paper and its antibacterial activity study. Green Chem. **8**(12), 1034–1041 (2006)
7. C. Liangyuan et al., Synthesis of ZnO–SnO_2 nanocomposites by microemulsion and sensing properties for NO_2. Sens. Actuators B: Chem. **134**, 360–366 (2008)
8. A. Kolmakov et al., Detection of CO and O^2 using tin oxide nanowire sensors. Adv. Mater. **15**(12), 997–1000 (2003)
9. O. Lupan et al., A rapid hydrothermal synthesis of rutile SnO_2 nanowires. Mater. Sci. Eng., B **157**(1), 101–104 (2009)
10. E. Comini et al., Stable and highly sensitive gas sensors based on semiconducting oxide nanobelts. Appl. Phys. Lett. **81**(10), 1869–1871 (2002)
11. P. Rai et al., Synthesis of flower-like ZnO microstructures for gas sensor applications. Sens. Actuators B: Chem. **178**, 107–112 (2013)
12. A. Kołodziejczak-Radzimska, T. Jesionowski, Zinc oxide—from synthesis to application: a review. Materials **7**(4), 2833–2881 (2014)
13. T. Han et al., Snowflake-shaped ZnO nanostructures-based gas sensor for sensitive detection of volatile organic compounds. Adv. Cond. Matter Phys. **2017**, 1–7 (2017)
14. J. Huang et al., Preparation of porous flower-shaped SnO_2 nanostructures and their gas-sensing property. Sens. Actuators B: Chem. **147**(2), 467–474 (2010)
15. S. Gnanam, V. Rajendran, Anionic, cationic and nonionic surfactants-assisted hydrothermal synthesis of tin oxide nanoparticles and their photoluminescence properties. Dig. J. Nanomater. Bios. **5**(2), 623–628 (2010)
16. M. Chitra et al., ZnO/SnO_2/Zn_2SnO_4 nanocomposite: preparation and characterization for gas sensing applications. Nanosystems: Phys. Chem. Math **7**(4), 707–710 (2016)
17. S. Thiagarajan, A. Sanmugam, D. Vikraman, *Facile methodology of sol-gel synthesis for metal oxide nanostructures in recent applications in sol–gel synthesis,* ed. by U. Chandra, InTechOpen, 1–16 (2017)
18. V. Sharma, R.C. Prajapati, Synthesis of mixed metal oxide nanoparticles of SnO_2 with SrO via sol–gel technology: their structural, optical, and electrical properties. J. Sol-Gel Sci. Technol. **87**(1), 41–49 (2018)

19. T.F. Choo, N.U. Saidin, K.Y. Kok, Hydrogen sensing enhancement of zinc oxide nanorods via voltage biasing. R. Soc. Open Sci. **5**(5), 172372 (2018)
20. D.R. Miller, S.A. Akbar, P.A. Morris, Nanoscale metal oxide-based heterojunctions for gas sensing: A review. Sens. Actuators B: Chem. **204**, 250–272 (2014)
21. B.D. Cullity, Elements of x-ray diffraction. Addison- Wesley Publishing Company Inc., Boston, 514 (1956)
22. S. Mohana Priya, A. Geetha, K. Ramamurthi, Structural, morphological and optical properties of tin oxide nanoparticles synthesized by sol–gel method adding hydrochloric acid. J. Sol-Gel Sci. Technol. **78**(2), 365–372 (2016)
23. X. Song et al., A highly sensitive ethanol sensor based on mesoporous ZnO-SnO2 nanofibers. Nanotechnology **20**(7), 075501–075501 (2009)
24. F. Gyger et al., Nanoscale SnO_2 hollow spheres and their application as a gas-sensing material. Chem. Mater. **22**(16), 4821–4827 (2010)
25. T. Sahoo et al., Nanocrystalline ZnO thin films by spin coating-pyrolysis method. J. Alloy. Compd. **491**(1–2), 308–313 (2010)
26. F. Gu et al., Synthesis and luminescence properties of SnO_2 nanoparticles. Chem. Phys. Lett. **372**, 451–454 (2003)
27. J.-C. Li, X.-Y. Hou, Q. Cao, Effect of Zn/Sn ratio on structure and properties of ZnO–SnO_2 nanocomposite films. J. Alloy. Compd. **611**, 219–224 (2014)
28. Y. Jiang et al., Unusual enhancement in electrical conductivity of tin oxide thin films with zinc doping. Phys. Chem. Chem. Phys. **13**(13), 5760–5763 (2011)
29. H. Xu et al., A novel method for improving the performance of ZnO gas sensors. Sens. Actuators B: Chem. **114**(1), 301–307 (2006)
30. M. Verma, V. Gupta, P1. 0.13 Highly sensitive ZnO−SnO_2 nanocomposite H_2 gas sensor. Proc. IMCS, 787–790, (2012)
31. B. Mondal et al., ZnO–SnO_2 based composite type gas sensor for selective hydrogen sensing. Sens. Actuators B: Chem. **194**, 389–396 (2014)

Chapter 6
Tunable Exchange Bias Behavior Near Room Temperature in Spinel Chromite

Junmoni Barman and S. Ravi

Abstract $Ni(Cr_{0.60}Fe_{0.40})_2O_4$ is prepared in single phase form by sol–gel method. Analysis of X-ray diffraction pattern reveals cubic spinel structure with $Fd\bar{3}m$ space group. As per magnetization measurements, temperature induced magnetization reversal is observed near room temperature with a compensation temperature of 366 K. Exchange bias behavior with tunable positive and negative exchange bias field is also observed in the vicinity of compensation temperature. These behaviors are explained by considering different temperature dependence of different sublattice magnetic moments and the competition between them. Furthermore, the bipolar switching of magnetization is demonstrated at room temperature.

6.1 Introduction

Magnetic spinel chromites are one of the most fascinating magnetic spinel family owing to their rich magnetic phase diagram and strong correlations among spin, charge and lattice degree of freedom. Such correlations give rise to magnetoelastic, magnetodielctric and magnetoelectric multifferoic properties in spinel chromites [1]. Existence of exchange bias (EB) and magnetization reversal (MR) behaviors in such multiferroic materials add to multi-functionality and enhance their potential for niche applications in spintronics, especially if these properties exist near room temperature. MR is an effect of switching the magnetization from positive to negative or in reverse without varying the applied magnetic field direction but simply by changing the magnitude of applied field. Switching of magnetization is also observed due to temperature variation [2]. Whereas, the exchange anisotropy created at the interface between a ferromagnet/ferrimagnet and an antiferromagnet originates the EB phenomenon which gives rise to magnetic hysteresis loop shift

J. Barman (✉)
Department of Physics, Rajiv Gandhi University, Arunachal Pradesh 791112, India

S. Ravi
Department of Physics, Indian Institute of Technology Guwahati, Guwahati 781039, India

R. G. Nair et al. (eds.), *Proceedings of 28th National Conference on Condensed Matter Physics*, Springer Proceedings in Physics 269,
https://doi.org/10.1007/978-981-16-5407-7_6

along the field axis [3]. In addition to conventional negative EB, coexistence of a positive and negative EB, i.e. tunable EB behavior is also observed in some materials. The simultaneity of MR and EB behavior were explored in wide variety of materials, for example in orthochromites $La_{1-x}Pr_xCrO_3$ [4], bulk single phase oxide materials Sr_2YbRuO_6 [5], $YFe_{0.5}Cr_{0.5}O_3$ [6], $La_{0.2}Ce_{0.8}CrO_3$ nanoparticles [7], rare earth intermetallic compounds $Nd_{0.75}Ho_{0.25}Al_2$ [8], spinel $Co(Cr_{0.95}Fe_{0.05})_2O_4$ [9], solid solutions of $BiFeO_3$–$BiMnO_3$ [10], etc. But to be profitable for practical applications, these behaviors should be observed near room temperature. However, as per our knowledge all the literatures available exhibit these behaviors well below the room temperature with very low magnetic compensation temperature (T_{comp}). So picking a material exhibiting both these behaviors with T_{comp} near room temperature is still a challenging problem.

$NiCr_2O_4$ is a normal cubic spinel oxide with $Fd\bar{3}m$ space group at temperature above 320 K. Due to Jahn–Teller distortion (JTD) at 320 K, the structure of $NiCr_2O_4$ transforms into tetragonal with $I4_1/amd$ space group below 320 K [11]. $NiCr_2O_4$ is composed of two sublattices at different crystallographic sites, i.e. at tetrahedral A-site and octahedral B- site. However, as per the magnetic structure of $NiCr_2O_4$ [12], all the A sites are collectively considered as one sublattice whereas B sites are split into two sublattices with each sublattice possessing both longitudinal and transverse components of magnetic moment. The A and B sites are occupied respectively by Jahn–Teller active Ni^{2+} ($e_g^4t_{2g}^4$) ions and Cr^{3+} ($t_{2g}^3e_g^0$) ions. Such complex magnetic structure brings about several exciting phenomena in $NiCr_2O_4$ including the EB [13, 14]. Substitution of other transition elements in different sublattice sites of $NiCr_2O_4$ may also highly influence its structural as well as magnetic properties giving rise to MR. The Fe^{3+} ($3d^5$) ions have a higher magnetic moment of 5 μ_B compared to that of Cr^{3+} ($3d^3$) ions (3 μ_B). Therefore, in the current work Fe substituted $NiCr_2O_4$ [$Ni(Cr_{0.60}Fe_{0.40})_2O_4$] compound is prepared expecting larger transition temperature (T_C) and enhanced T_{comp}. We demonstrated both MR and tunable EB behavior near room temperature; along with the room temperature bipolar switching of magnetization.

6.2 Experimental Techniques

Polycrystalline $Ni(Cr_{0.60}Fe_{0.40})_2O_4$ was prepared by sol–gel technique, starting from the stoichiometric proportion of $Ni(NO_3)_2.6H_2O$, $Fe(NO_3)_3.9H_2O$ and $Cr(NO_3)_3.9H_2O$. These compounds, as per stoichiometric ratio are dissolved in distilled water and then mixed together. The resultant mixture is then kept under constant magnetic stirring at temperature of around 80–100 °C for 3–4 h by placing it in a hot plate and, proper amount of citric acid and ethylene glycol are added. Finally, a metal citrate gel is obtained. The gel is then fired at high temperature to obtain a fine powder. The precursor powder thus obtained was presintered at 600, 800 and 1000 °C for 12 h followed by final sintering at 1100 °C for 24 h. X-ray diffraction (XRD) pattern was obtained at Room temperature using a Rigaku make TTRAX III

X-ray diffractometer employing Cu-Kα ($\lambda = 1.54056$ Å) radiation. Microstructure and composition of the sample were examined using ZEISS make FESEM (ΣIGMA) furnished with EDS facility. Magnetic measurements were done using a Lakeshore make VSM of model no. 7410.

6.3 Results and Discussions

6.3.1 *Structural Properties*

Rietveld refinement of the XRD pattern verifies the phase purity of $Ni(Cr_{0.60}Fe_{0.40})_2O_4$. The observed XRD pattern together with the calculated data, obtained by applying the Rietveld refinement method are shown in Fig. 6.1. Refinement shows that $Ni(Cr_{0.60}Fe_{0.40})_2O_4$ exhibits cubic spinel structure at room temperature with $Fd\overline{3}m$ space group and the obtained lattice parameter is $a = 8.3012$ Å. On the other hand, at room temperature the parent $NiCr_2O_4$ is known to exhibit a tetragonal structure with $I4_1/amd$ space group [13]. Thus Fe doping reduces the JTD in parent $NiCr_2O_4$ thereby causing a structural transformation from tetragonal to cubic. Such reduction in JTD is because of the substitution of a few Fe^{3+} ions at the A site also, instead of residing only the B site. An equivalent amount of Ni^{2+} ions are therefore likely to be present in the B site.

Figure 6.2a shows the microstructure of the sample. The grain sizes of the sample vary over a broad range (Fig. 6.2b). The average grain size is estimated to be 390 nm with a standard deviation of 0.16 nm by fitting the size distribution histogram (Fig. 6.2b) to the log normal distribution function [15],

$$f(d, \mu, \sigma) = \frac{1}{d\sigma\sqrt{2\pi}} \exp\left(-\frac{(\ln d - \mu)^2}{2\sigma^2}\right)$$

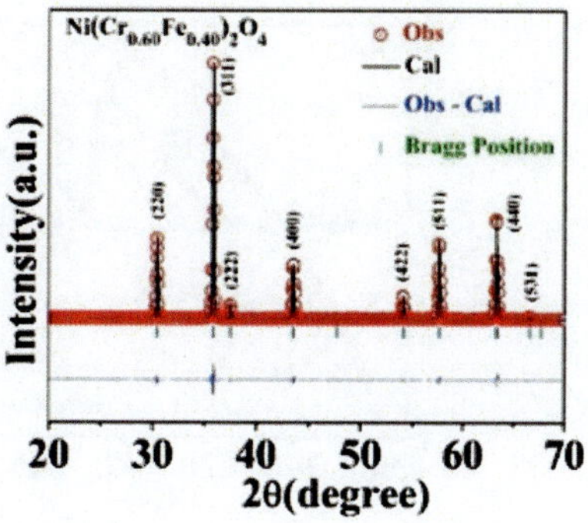

Fig. 6.1 XRD pattern of $Ni(Cr_{0.60}Fe_{0.40})_2O_4$ together with the calculated data as per Rietveld refinement

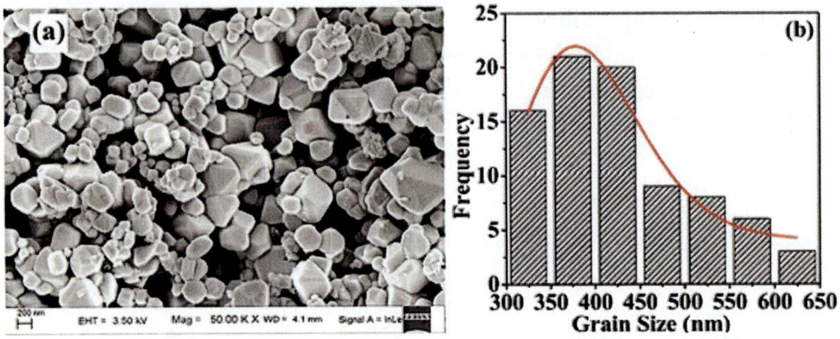

Fig. 6.2 **a** FESEM image and **b** grain size distribution of $Ni(Cr_{0.60}Fe_{0.40})_2O_4$

where, d, μ and σ represent the cross-sectional length of the particle, the logarithmic mean and the standard deviation respectively. As per the EDS analysis the chemical composition of the sample is comparable to that of the starting composition.

6.3.2 *Magnetization Reversal*

Temperature dependent magnetization measurement (*M–T*) was carried out under both zero field cooled (ZFC) and field cooled (FC) conditions by applying a magnetic field $H = 200$ Oe (Fig. 6.3a). A peak observed in the ZFC plot highlights the ferrimagnetic (FIM) transition. The T_C of the sample is found to be 504 K which is relatively larger than 73 K of parent $NiCr_2O_4$ [13].

The FC curve in Fig. 6.3a clearly depicts the occurrence of temperature induced MR in the sample with magnetic compensation close to room temperature, i.e., $T_{comp} = 366$ K. However, T_{comp} is found to decrease slightly when the applied magnetic field is increased. Eventually the MR vanishes completely for $H > 1.5$ kOe and

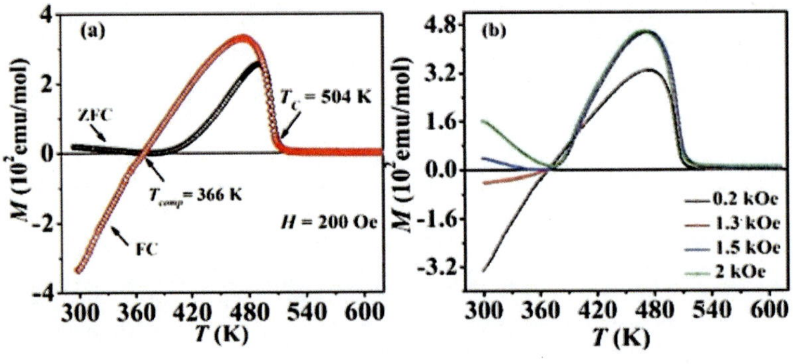

Fig. 6.3 **a** ZFC and FC *M–T* plots for $H = 200$ Oe and **b** FC *M–T* plots for different H

only positive magnetization is observed down to 300 K (Fig. 6.3b). The occurrence of MR in the current sample can be regarded as arising from different temperature dependences of the net A and B sublattice magnetic moments as follows: for $T_{comp} < T < T_C$, the net A (Ni + Fe) sublattice moment lies in the applied field direction and dominates over the net B (Cr1 + Cr2 + Fe + Ni) sublattice moment which lies in opposite direction to that of A, i.e., here $\mu_A > \mu_B$ and therefore results in a net positive magnetization. When the temperature is decreased, the net B sublattice moment start increasing slowly and compensates the A sublattice moment at $T = T_{comp}$, thereby resulting a net zero magnetic moment ($M = 0$) at T_{comp}. If the temperature is dropped further, the net B sublattice moment keep on increasing and finally $\mu_A < \mu_B$, leading to a net negative magnetization. In order to avoid any intricacy in understanding, in the present discussion we have just considered the longitudinal components and excluded the transverse components of magnetic moments. As per earlier reports, no magnetization reversal behavior is observed in parent $NiCr_2O_4$ where net A sublattice moment dominates over net B sublattice moment till low temperature (saturation magnetization value of 0.3 μ_B/f.u. [12]). Upon Fe substitution, the B sublattice moment increases and both A and B sublattice moments become comparable. Hence, net B sublattice moment is likely to dominate the A sublattice moment for $T < T_{comp}$, resulting in sign reversal of magnetization.

6.3.3 Exchange Bias

For the investigation of EB effect, the sample was cooled under $H = 3000$ Oe from $T > T_C$ to the particular temperature every time and hysteresis (*M–H*) loops were traced. The *M–H* loops traced at $T < T_{comp}$ are observed to shift along positive field axis indicating positive EB behavior. Then the loop gradually shifts towards the origin with increase in temperature and at $T = T_{comp}$, it becomes symmetric about the origin. Thus no EB is observed at $T = T_{comp}$. When the temperature is increased beyond T_{comp}, the loops shift along the negative field axis indicating the negative EB behavior. This behavior of the sample at some selected temperatures is illustrated in Fig. 6.4. At $T = T_{comp}$ (=366 K), symmetric loop with almost linear behavior or with minimum coercivity is observed. However, while increasing or decreasing the temperature on either side of T_{comp}, opening in the loop is observed. This behavior implies that close to T_{comp}, A and B sublattice moments are almost compensated. Shifting of the loops is significant only within 348 K < T < 383 K and elsewhere usual symmetric FIM loops are observed. As shown in Fig. 6.5a, the exchange bias field (H_{EB}) of the present sample first rises with rise in temperature and attains a maximum value of $H_{EB} = 1.59$ kOe at 361 K, and then it changes sign across T_{comp} and attains a maximum value of $H_{EB} = -1.97$ kOe at 370 K. The variation of effective coercive field (H_C^{eff}) with temperature shows a broad twin peak behavior (Fig. 6.5b) with minimum value in the close proximity of T_{comp} due to the compensated moments of A and B site.

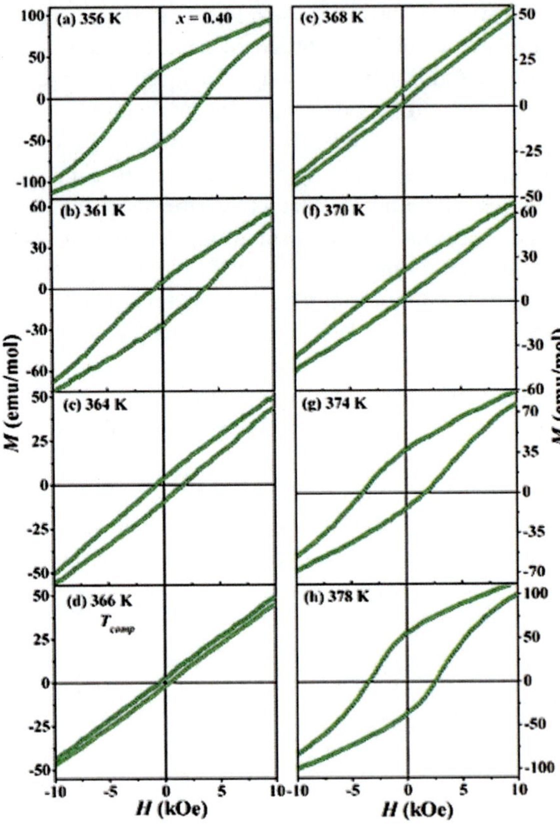

Fig. 6.4 FC ($H = 3000$ Oe) M–H loops close to T_{comp} in an expanded scale

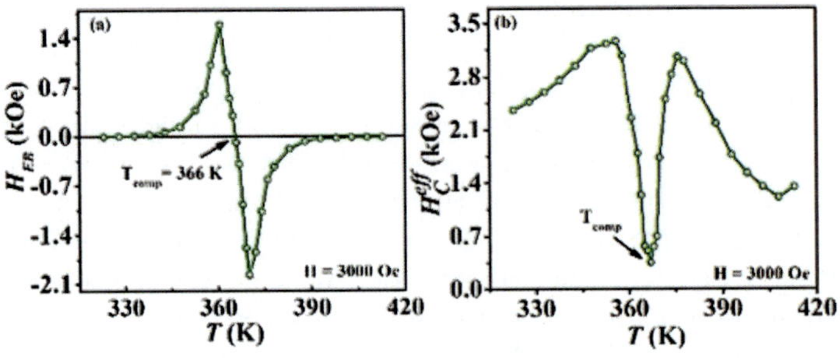

Fig. 6.5 Variation of **a** H_{EB} and **b** $H_C{}^{eff}$ for $Ni(Cr_{0.60}Fe_{0.40})_2O_4$ with temperature

The existence of tunable positive and negative EB behavior in $Ni(Cr_{0.60}Fe_{0.40})_2O_4$ can be explained by taking into account the change in domination of one FIM sublattice moment over the other with temperature variation. As discussed in the above section, for $T_{comp} < T < T_C$, the moment of A sublattice is higher than that of B, i.e.$\mu_A > \mu_B$ and here μ_A lies in the direction of applied field. Because of the dominant μ_A in the positive field direction, this circumstance promotes easy alignment of the net magnetization in the applied field direction while recording the ascending branches of the hysteresis loop. However, when the field is applied in the reverse direction to record the descending branch, now it is comparatively difficult to orient μ_A in the field direction from its ground state direction. Therefore, a very high negative field is needed to align the net magnetization in the field direction. Consequently, as a whole the hysteresis loop shifts along the negative field axis, *i.e.*, they manifest negative EB field. On the other hand, in the temperature range $T < T_{comp}$, $\mu_B > \mu_A$ and μ_B aligns along negative field direction, so here comparatively strong positive field is necessary to line up the net magnetization while recording the ascending branches of the loop. Hence in this range the loop as a whole shift along the positive field axis and positive EB field is achieved. At $T = T_{comp}$, since $\mu_A = \mu_B$, the EB is almost zero and a symmetric *M*–*H* loop is observed.

6.3.4 Bipolar Switching of Magnetization

The magnetic field induced MR behavior is studied at room temperature. Such field induced MR is also termed as bipolar switching of magnetization. For this purpose, $Ni(Cr_{0.60}Fe_{0.40})_2O_4$ was cooled under $H = 200$ Oe from $T > T_C$ to room temperature. The magnetization was then recorded at room temperature for the same field for 200 sec and it is found to be negative. Afterward, the applied magnetic field is increased to 1850 Oe and a positive magnetization is observed. In this way the magnetic field is cycled between +200 Oe and +1850 Oe for a number of times and a simultaneous negative and positive magnetization is observed. Thus in addition to temperature induced MR, the bipolar switching of magnetization is also possible in the present sample (Fig. 6.6). But after the completion of the first cycle, the

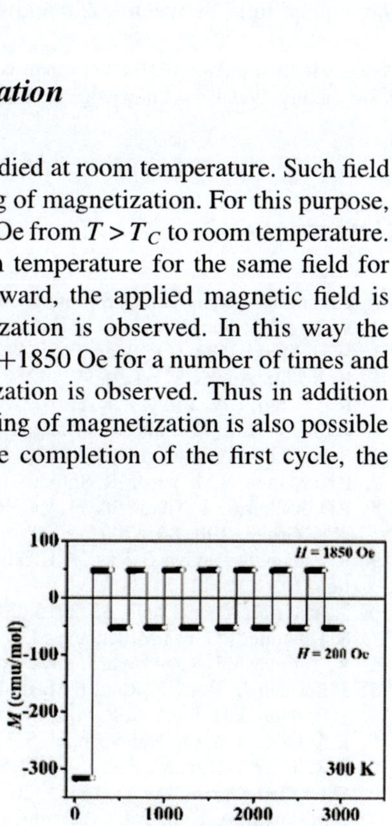

Fig. 6.6 Bipolar switching of magnetization at room temperature

negative magnetization was found to be reduced from its initial value. However, for the subsequent cycles, magnitude of negative magnetization remains constant. The magnitude of magnetization in this reversible part is higher than that of $YFe_{0.5}Cr_{0.5}O_3$ [16], observed by cycling the field between +50 Oe and +1200 Oe at 200 K. This behavior was noticed in wide variety of materials [17, 18] but most of them are at far below the room temperature.

6.4 Conclusions

In conclusion, simultaneity of magnetization reversal and EB field reversal is observed in the close proximity of room temperature in $Ni(Cr_{0.60}Fe_{0.40})_2O_4$, prepared by sol-gel route. The MR in the present sample is explained by taking into account different temperature dependence of the different sublattice moments. Tunable EB field is also explained fruitfully by taking into account the change in domination of one sublattice moment over the other because of the variation in temperature. Room temperature bipolar switching of magnetization was demonstrated simply by cycling the applied field between +200 and +1850 Oe.

Acknowledgements CIF, IIT Guwahati is acknowledged for the FESEM and high temperature VSM facility. We also acknowledge DST-FIST for XRD facility.

References

1. K. Singh, A. Maignan, C. Simon, C. Martin, Appl. Phys. Lett. **99**, 172903 (2011)
2. R. Ang, Y.P. Sun, X. Luo, C.Y. Hao, X.B. Zhu, W.H. Song, J. Appl. Phys. **104**, 023914 (2008)
3. W.H. Meiklejohn, C.P. Bean, Phys. Rev. **102**, 1413 (1956)
4. K. Yoshii, Appl. Phys. Lett. **99**, 142501 (2011)
5. R.P. Singh, C.V. Tomy, A.K. Grover, Appl. Phys. Lett. **97**, 182505 (2010)
6. J. Mao, Y. Sui, X. Zhang, X. Wang, Y. Su, Z. Liu, Y. Wang, R. Zhu, Y. Wang, W. Liu, X. Liu, Solid State Commun. **151**, 1982 (2011)
7. P.K. Manna, S.M. Yusuf, R. Shukla, A.K. Tyagi, Appl. Phys. Lett. **96**, 242508 (2010)
8. P.D. Kulkarni, A. Thamizhavel, V.C. Rakhecha, A.K. Nigam, P.L. Paulose, S. Ramakrishnan, A.K. Grover, EPL **86**, 47003 (2009)
9. R. Padam, S. Pandya, S. Ravi, A.K. Nigam, S. Ramakrishnan, A.K. Grover, D. Pal, Appl. Phys. Lett. **102**, 112412 (2013)
10. A.A. Belik, Inorg. Chem. **52**, 2015 (2013)
11. S. Klemme, J.C. van Miltenburg , Phys. Chem. Minerals **29**, 663 (2002)
12. K. Tomiyasu, I. Kagomiya, J. Phys. Soc. Jpn. **73**, 2539 (2004)
13. J. Barman, T. Bora, S. Ravi, J. Magn. Magn. Mater. **385**, 93 (2015)
14. J. Barman, P.D. Babu, S. Ravi, J. Magn. Magn. Mater. **418**, 300 (2016)
15. E.A. Odo, Nanosci. Nanotechnol. **5**, 57 (2015)
16. L.R. Shi, Z.C. Xia, M. Wei, Z. Jin, C. Shang, J.W. Huang, B.R. Chen, Z.W. Ouyang, S. Huang, G.L. Xiao, Ceram. Int. **41**, 13455 (2015)
17. S.M. Yusuf, A. Kumar, J.V. Yakhmi, J. Phys.:Conf. Ser. **200**, 022073 (2010)
18. P. Mandal, A. Sundaresan, C.N.R. Rao, A. Iyo, P.M. Shirage, Y. Tanaka, C. Simon, V. Pralong, O.I. Lebedev, V. Caignaert, B. Raveau, Phys. Rev. B **82**, 100416 (2010)

Chapter 7
Performance Engineering of C-Doped Titania for Photocatalytic Hydrogen Production Through pH Tuning

S. K. Nikhil, Joshi Pushkar Shrikant, and Ranjith G. Nair

Abstract Titania (TiO_2) based photocatalytic water spitting has excellent potential for hydrogen production. However, the poor light-harvesting capacity and high recombination rate of photoinduced electron–hole pairs limit its performance. Therefore, doping of TiO2 with metal/non-metal ions can be a practical approach to improve the performance of TiO_2 for efficient photocatalytic H_2 production. Along with the role of dopants, various synthesis parameters also contribute to the performance of TiO_2. Considering this, the present study has shown the influence of pH in the performance of C-doped TiO_2. C-doped TiO2 under various pH values (3,7,11) was synthesised via the sol–gel method. The structural morphological and optical characterisation was performed using XRD, FESEM and UV–Vis spectroscopy. The C doping was confirmed through EDS. The samples were further tested for hydrogen production. It was observed that the C-dopedTiO_2 synthesised under pH 7 produced hydrogen gas at a rate of 1142.4 μmol/g/h, which is about 3 times and 1.5 times greater than that produced by samples synthesised under pH 3 and 11 conditions. The superior performance of the photocatalyst at the optimum pH value indicates the role of pH tuning on the performance of C doped TiO_2 for efficient H_2 production.

7.1 Introduction

Efficient energy usage is the key to sustaining the economy, conserve the environment, and the future of humanity. The economic and environmental impact of fossil fuel has diverted the research for utilising hydrogen and other conventional energy sources. Among these, hydrogen is being projected as a potential candidate because of the several advantages it possesses, like combustion that does not produce toxic gases and can be stored as a renewable energy source [1, 2]. Among the different techniques for hydrogen production, production via solar photocatalysis is a promising choice considering the environmental impact and storage requirements [3].

S. K. Nikhil · J. P. Shrikant · R. G. Nair (✉)
Solar Energy Materials Research & Testing Laboratory (SMaRT Lab), Department of Physics, National Institute of Technology Silchar, Silchar 788010, Assam, India

R. G. Nair et al. (eds.), *Proceedings of 28th National Conference on Condensed Matter Physics*, Springer Proceedings in Physics 269,
https://doi.org/10.1007/978-981-16-5407-7_7

Among the different metal oxide semiconductors considered as a photocatalyst, TiO_2 shows a better performance. Also, the low toxic nature and low cost make TiO_2 preferable for practical application. However, the activity of TiO_2 is limited to the UV region of the solar spectrum due to its wide bandgap nature [4]. A faster recombination rate also hinders the performance of TiO_2. Various strategies have been adopted for performance enhancement, like doping with metals/non-metals. Along with that, engineering the synthesis conditions such as variation in pH has also been found to influence the surface properties of TiO_2 [5].

The present work reports the synthesis of carbon-doped TiO_2 under various pH conditions using a simple sol gel route. The synthesised samples were tested for hydrogen production application. The influence of pH condition on the formation of TiO_2 and thus its performance are studied.

7.2 Experimental

7.2.1 Materials and Methods

Titanium isopropoxide ($Ti[OCH (CH_3)]_4$) and D-Glucose powder ($C_6H_{12}O_6$) were the precursors used for titanium and carbon, respectively. Sodium hydroxide (NaOH) pellets and 90% glacial acetic acid (CH_3COOH) were used to make the acidic and basic buffers, respectively. Ethanol was used as the solvent throughout the synthesis. Titanium isopropoxide, glacial acetic acid, and EMSURE grade ethanol were all obtained from Sigma-Aldrich, India. NaOH pellets and D-Glucose powder were procured from Merck, India. Only double-distilled water was used throughout the process and all the chemicals used were ensured to be of the highest purity.

In the sol–gel synthesis of carbon-doped TiO_2, a simple procedure was followed. Firstly, 90% glacial acetic acid solution and NaOH pellets were dissolved in different beakers containing ethanol in certain amounts to make the acidic and basic buffer solutions. Then, these buffers were dissolved in appropriate amounts in three beakers containing ethanol to get the pH values of 3, 7, and 11. Subsequently, 1 ml of Titanium (IV) isopropoxide (TTIP) was added dropwise to each beaker to get clear solutions. These solutions were stirred at 300 rpm at room temperature. After an hour, an equal amount of double-distilled water was added to each of the solutions dropwise for hydrolysis. Along with the water, the glucose powder was introduced to each beaker in quantities determined by molar calculations. The resultant-milky sol was stirred at 300 rpm for further 24 h at room temperature and subsequently allowed to age further 12 h after the stirrer was turned off. Then, the sol was dried in a hot air oven at 80 °C for 12 h to dry off excess amount of ethanol. The so-formed flaky powders with the starting pH 3, 7, and11 were labelled TC3, TC7, and TC11, respectively. These powders were then calcined at 650 ˚C.

7.2.2 Characterisation

The X-ray diffraction analysis was done using Panalytical XPERT[3] Powder diffractometer. Scanning Electron Microscopy with Energy Dispersive X-ray spectroscopy characterisation was carried out using ZEISS ultra field emission SEM. The optical characterisation studies was carried out using CARY 5000(Agilent Technologies) with DRS mode.

7.2.3 Hydrogen Production

Hydrogen production ability of the carbon doped TiO2 samples of solvent pH values 3, 7, and 11 calcined at 650 °C for 1 h was tested using the Agilent—490 Micro GC system under ultraviolet irradiation. The following procedure was followed in preparing the samples for testing. 20 mg of each sample was mixed in a round bottom flask of a known volume with 35 ml distilled water. The round bottom flask was sealed airtight and subjected to sonication for 10 min to ensure uniform dispersal of the sample. 15 ml of pure methanol (CH_3OH) was then added to the flask as a sacrificing agent. It was also loaded with 20 μl of chloroplatinic acid (H_2PtCl_6) solution coat the TiO_2 surface with platinum. The flask was sealed airtight again. The results of these hydrogen production capability tests of the carbon-doped TiO_2samples are presented ahead.

7.3 Result and Discussion

The XRD patterns of the samples are given in Fig. 7.1. The XRD pattern confirms the crystalline structure of the synthesised samples. All the samples showed pure anatase peak with no traces of rutile/brookite peaks. The average crystallite sizes of the samples were determined using Scherrer's equation and are given in Table 7.1. The increase in crystallite size with the increase in pH is due to an increase in microstrain [6].

Figure 7.2a–c shows the FESEM images of the synthesised TiO_2 samples. The TC3 sample does exhibits a flake-like structure. Spherical shaped particles are formed in the case of TC7 and TC11 samples. It can also be seen that in the case of TC7 and TC11, particles have agglomerated as result comparatively high calcination temperature [7]. The carbon doping is further confirmed by the EDS spectra, as shown in Fig. 7.3.

Figure 7.4a–c shows the UV–Vis absorption spectra of the synthesised samples along with the respective Tauc's plot [8]. All the samples exhibit intense absorption in the UV region. The band gap for the samples are calculated and given in Table 7.1.

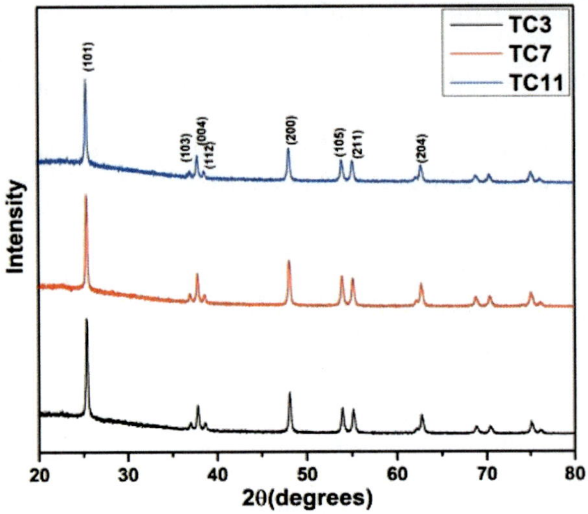

Fig. 7.1 XRD pattern of the samples

Table 7.1 Various characterisation parameters from the studies conducted

Sample	Crystallite Size (nm)	Bandgap (eV)	H_2 in 120 min (μmol/g)	H_2 in 120 min (μmol/g/h)
TC3	38.25	3.28	741.60	370.80
TC7	38.58	3.01	2284.97	1142.48
TC11	39.69	3.17	1442.98	721.49

The effect of pH on the band gap is visible with the sample TC7 having the lowest bandgap among the three samples.

The hydrogen production using the TiO_2 samples were done under standard conditions. The rate of hydrogen evolution was plotted against time and is shown in Fig. 7.5 and the values are given in Table 7.1. The TC7 sample outperformed both the other samples. In 2 h of photocatalysis reaction, the sample TC7 produced 2.285 ml of hydrogen gas per unit ml of water at the rate of 1.143 ml of H_2 per ml of water per hour.

The superior performance of TC7 over other samples can be attributed to lower bandgap energy, allowing better light-harvesting capacity.

7.4 Conclusion

Carbon doped TiO_2 was synthesised under different pH conditions. The XRD pattern shows phase stabilisation with no visible rutile/brookite peaks. The FESEM analyses show the change in surface morphology with the change in pH conditions, with particles getting spherically agglomerated with increase in pH. The agglomeration may

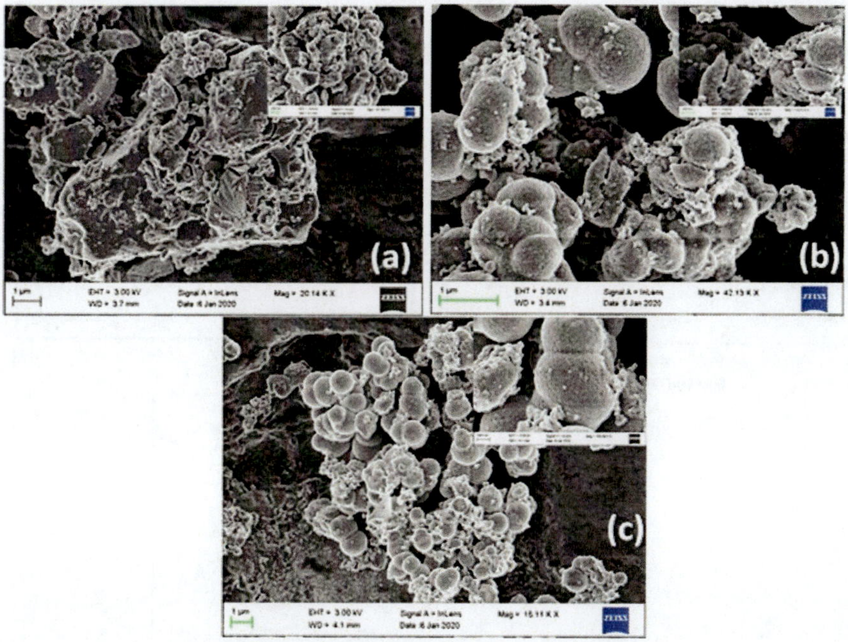

Fig. 7.2 FESEM images of **a** TC3, **b** TC7 and **c** TC11

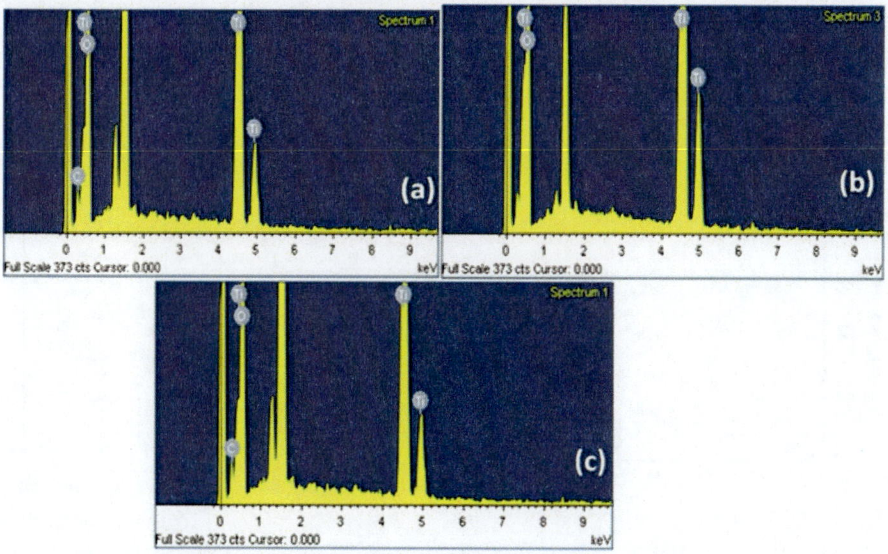

Fig. 7.3 EDS spectra of **a** TC3, **b** TC7 and **c** TC11

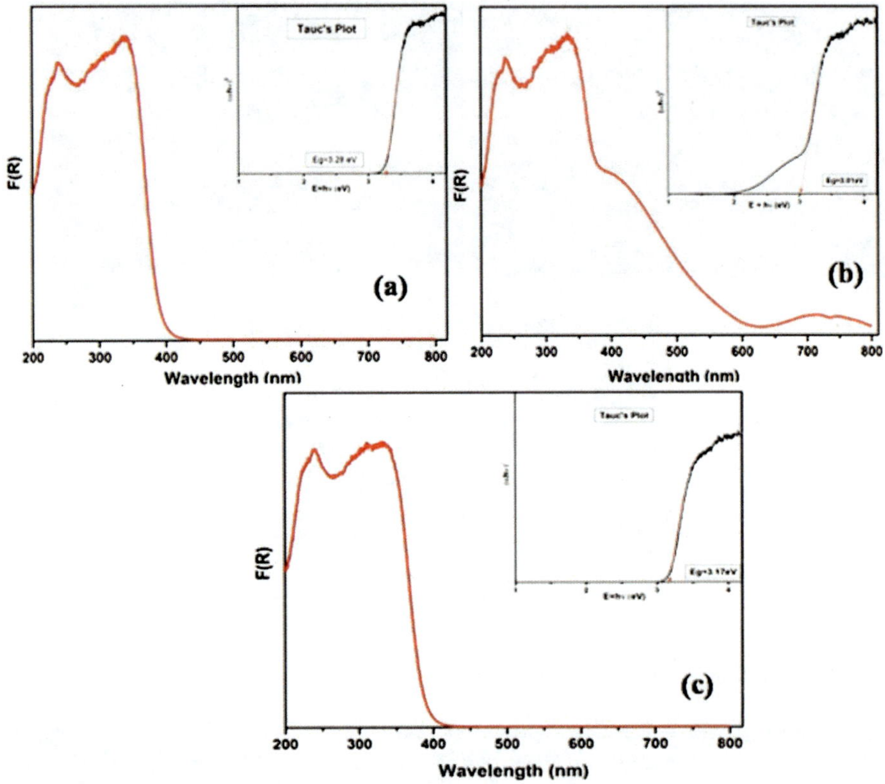

Fig. 7.4 UV–Vis spectra of **a** TC3, **b** TC7 and **c** TC11 with Tauc's Plot

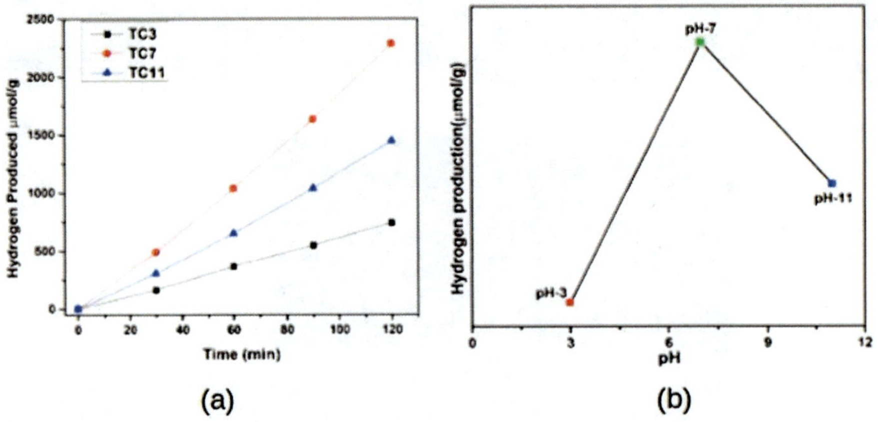

Fig. 7.5 **a** Rate of hydrogen evolution, **b** Hydrogen production vs pH

also be due to the high calcination temperature. Optical studies showed a decrease in band gap energy in TC7 sample as compared to the other two samples. The reduction of the band gap energy in the case of TC7 sample led to a better performance of the same in the hydrogen production.

Acknowledgements The authors express their sincere gratitude to Prof. Vivek Polishettiwar, Department of Chemical Sciences,Tata Institute of Fundamental Research, Mumbai for allowing to conduct characterisation and hydrogen production experiment in his lab.

References

1. M. Balat, Potential importance of hydrogen as a future solution to environmental and transportation problems. Int. J. Hydrogen Energy. **33**(15), 4013–4029 (2008)
2. N.Z. Muradov, T.N. Veziroğlu, "Green" path from fossil-based to hydrogen economy: An overview of carbon-neutral technologies. Int. J. Hydrogen Energy. **33**(23), 6804–6839 (2008)
3. N.S. Ibrahim, W.L. Leaw, D. Mohamad, S.H. Alias, H. Nur, A critical review of metal-doped TiO2 and its structure–physical properties–photocatalytic activity relationship in hydrogen production. Int. J. Hydrogen Energy [Internet]. **45**(53), 28553–28565 (2020)
4. H. Dong, G. Zeng, L. Tang, C. Fan, C. Zhang, X. He et al., An overview on limitations of TiO2-based particles for photocatalytic degradation of organic pollutants and the corresponding countermeasures. Water Res. **79**, 128–146 (2015)
5. J. Yuenyongsuwan, N. Nithiyakorn, P. Sabkird, E.A. O'Rear, T. Pongprayoon, Surfactant effect on phase-controlled synthesis and photocatalyst property of TiO2 nanoparticles. Mater Chem. Phys. **214**, 330–336 (2018)
6. A. Molea, V. Popescu, N.A. Rowson, A.M. Dinescu, Influence of pH on the formulation of TiO2 nano-crystalline powders with high photocatalytic activity. Powder Technol. **253**, 22–28 (2014)
7. D.J. Kim, S.H. Hahn, S.H. Oh, E.J. Kim, Influence of calcination temperature on structural and optical properties of TiO2 thin films prepared by sol-gel dip coating. Mater Lett. **57**(2), 355–360 (2002)
8. B. Choudhury, A. Choudhury, Oxygen defect dependent variation of band gap, Urbach energy and luminescence property of anatase, anatase-rutile mixed phase and of rutile phases of TiO2 nanoparticles. Phys E Low-Dimensional Syst. Nanostruct. [Internet]. **56**, 364–371 (2014)

Chapter 8
Development of Electrode Plates Using Vapour Deposition Method for RPC Detectors

Hemen Ch. Medhi and P. K. Boruah

Abstract Resistive plate chambers (RPCs) are rugged and affordable gaseous detectors that have found wide application in High-Energy Physics and astroparticle experiments. The main features of these detectors are the very large pulse height, reduced cost per unit area of coverage and good time resolution approximately 1 ns. The ease of design makes its application not only in the detection of charge particles but medical imaging also. RPCs are designed using a constant and uniform electric field on parallel electrode plates which are made of a material with high volume resistivity of the order of 10^{10}–10^{12} Ω cm. The electrodes used in this experiment is float glass, Al is deposited on the glass plate to make them conductive electrode plates. This paper deals with the design of glass electrode based single gap, glass RPC prototype of 15 cm × 15 cm size, using vapour deposition method and its testing using front end electronics.

8.1 Introduction

The Resistive plate chambers (RPCs) were introduced in 1981 by R. Santonico and R Cardreli, are gaseous detectors [1], based on the principle of "spark chamber" , using a constant and uniform electric field [2] produced by two parallel electrode plates which are made up of a material with high volume resistivity of the order of 10^{10}–10^{12} ohm-cm [3]. RPCs are based on the principle of gaseous ionization produced by charged particles traversing the active area of the detector, large avalanche of electron under a strong uniform electric field. A gas mixture of P10 and Freon [4], which is ionized by charged particles, is flown through the gap between the electrode plates. The electrodes used in this experiment are glass electrodes, deposited with a layer of conductor at outer side of the glass plate, using vapour deposition method.

H. Ch. Medhi (✉)
Department of Electronics St, Edmund's College, Shillong 793003, Meghalaya, India

P. K. Boruah
Department of Instrumentation, Gauhati University, Guwahati 781014, Assam, India

R. G. Nair et al. (eds.), *Proceedings of 28th National Conference on Condensed Matter Physics*, Springer Proceedings in Physics 269,
https://doi.org/10.1007/978-981-16-5407-7_8

8.2 Development of RPC

RPC can be developed and tested easily, which is very common for the other detectors in high energy physics experiments [5]. The main components used for the development of these detectors are a gas gap, pick up strips, electrode plates and front end electronics. The following steps are important for the development of RPC detector.

8.2.1 Cutting and Cleaning of Glass Plates and Spacers

We used 15 cm $\times$ 15 cm and thickness 2 mm, commercially available float glass to fabricate the resistive electrodes of the RPC. The glasses are cut to the required size with the help of a diamond cutter. The glass sheets are cleaned thoroughly by mild acid (chromic/acetic acid) and mild alkali wash. After completing each washing the plates are rinsed thoroughly with distilled water. The glass plates are then left for drying in ambient condition in a dust free polythene cabinet. The next step is to introduce the 2 mm thick window glass used for spacers are also cleaned with the same procedure.

8.2.2 Conductive Coating

The resistive coating of the outer surfaces of the electrodes plays a very crucial role in the operation of the RPC detector. While the surface resistivity of this coating should be small enough so that the bias voltage that is required for the RPC operation can be applied on these coats, it should be high enough to render it transparent to electric pulses generated by the charge displacement in the gas gap. This way, the charges produced inside the gas gap on passage of a particle, can induce electric signals on the external metallic pickup strips, which are capacitively coupled to the gas gap.

Several types of materials and variety of application methods for obtaining this coating can be found. A silk scrin printing method can also be used by which the coat can be controlled precisely. Other technique such as metal oxide paints, anti-static paints, graphite varnishes, adhesive graphite foils, special inks and commercial automobile paints are also used.

8.2.3 Vapour Deposition Method

In this method aluminium coating was developed on a glass sheet by vapour deposition method (vacuum Coater, Model No. VC112, Vacuum Instrument Company,

New Delhi), at first window glass sheet was cut to size 15 cm × 15 cm and then properly cleaned. The cleaning procedure involved cleaning the plates in chromic acid followed by rinsing in distilled water. The plates are dried and cleaned with alcohol. The clean plates are placed in the chamber of the vacuum coater. At temperature above boiling point of Aluminum and pressure 10^{-6} torr, Al vapour is formed which produces a fine coating of aluminum on exposed surface of the plates. The normal Ag coating on mirror glass and electro deposited Cu coating on phenolic materials such as printed circuit boards were also used as to act as conductive electrode where the resistive materials are either glass (in case of mirror) or phenolic (in case of PCB).

8.2.4 RPC Gas Chamber Design

All the glass components are finally cleaned with alcohol and rubbed with lint free cloth. Glass strips 1.25 cm wide and 2 mm thick are used to make the gas chamber frame and spacer in between the plates. A small drop of glue binds the spacer with one sheet of glass. The other glass plate is placed on top of the spacer uniformly, so that the glue is not spilled and glass sheets neatly in line with the one under it. The glue is then applied to the gap between the spacer and glass. It is important that the insulating glasses are arranged perfectly on each side of the assembly so that they will provide rigidity and mechanical strength required for the assembly to overcome the pressure exerted by the gas flowing through the whole chamber. The whole assembly is then left there for 24 h and become ready for leak test.

8.2.5 Design of Signal Pick Up Strip

Signal from the RPC is obtained from pick up strips, placed over the anode with proper insulation. The main criteria for selecting size of the strips and their characteristic impedance, cross talk and detector size. The pickup strips are laid on the pickup electrode plate. The strips are 3 cm wide with lengths covering the RPC active area and the separation between the strips is kept at 2 mm. Figure 8.1. shows a photograph of a complete RPC assembly.

8.3 RPC Efficiency Test

The final test of the RPC is done by measuring the efficiency of the RPC, measured at different bias voltages and gas ratios [6]. The schematic representation of the experimental set up for measuring detector efficiency is shown in Fig. 8.2. The RPC module, developed by vapour deposition method is sandwiched between the two scintillator (SC1 and SC2) each of dimension of 50 cm × 50 cm x 5 cm. To obtain

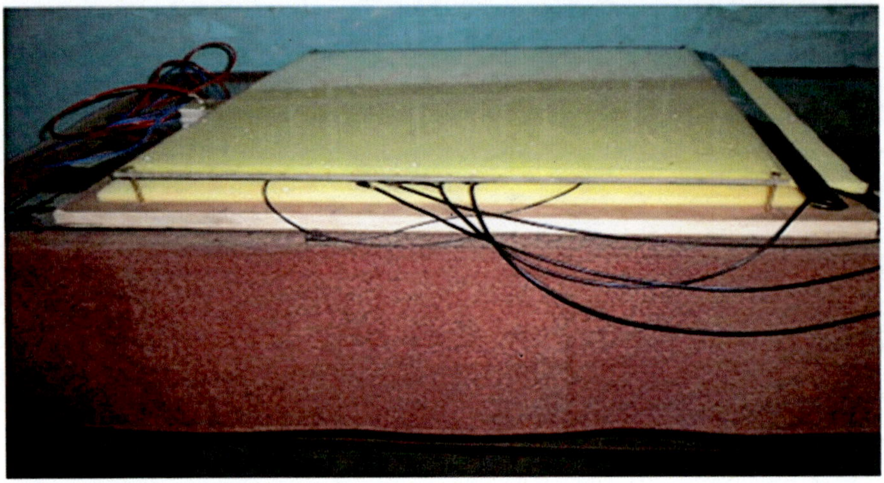

Fig. 8.1 Final Assembly of the RPC Detector

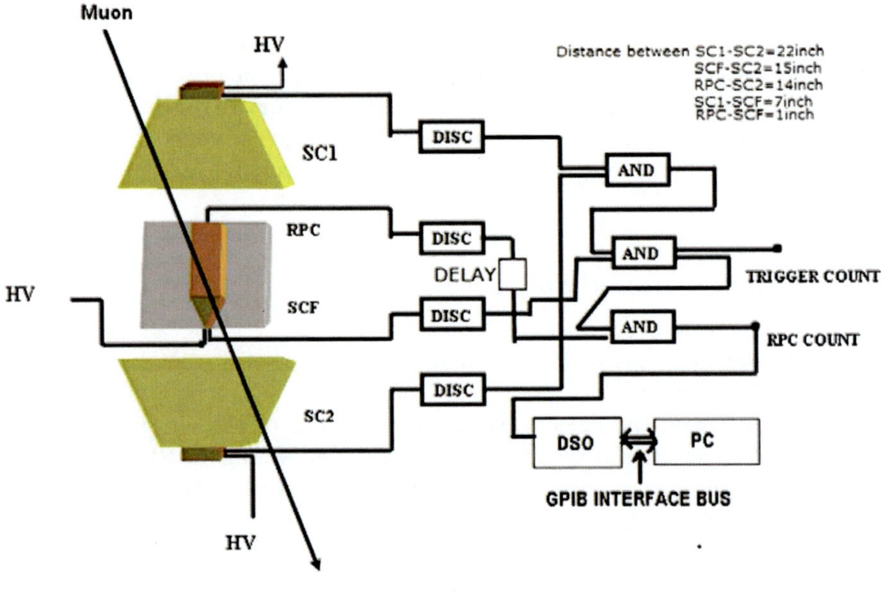

Fig. 8.2 Experimental set up for efficiency test for RPC [7]

the RPC efficiency in a region within one pickup strip, the trigger setup is further zoomed into a region of Finger scintillator (SCF) of dimension 5 cm × 5 cm × 5 cm, placed above the pickup strip. The trigger signal is obtained as SC1 AND SC2 AND SCF indicating passage of a cosmic ray muon. We therefore, obtain the efficiency in a region of 3 cm × 5 cm within a pickup strip. Coincidence between RPC count

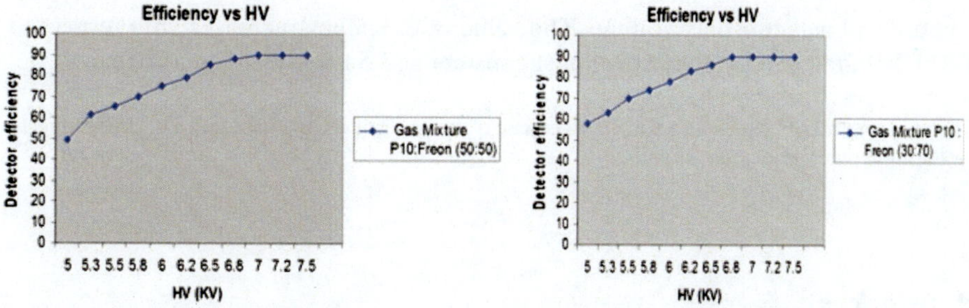

(a) Efficiency at 50:50 (b) Efficiency at 30:70

Fig. 8.3 Variation of Efficiency with Voltage (KV) for 15 cm × 15 cm, size RPC at different gas mixture

and trigger count is taken as final count. Therefore, efficiency is obtained from the following formulae.

$$\%\ Efficiency = \frac{RPC\ count\ with\ the\ signal\ in\ coincidence\ with\ trigger}{Trigger\ Count} \times 100$$

The presence of a muon trajectory and efficiency of the RPC is determined by the fraction of events, where RPC signal is above the threshold value of the discriminator. The pulses from the 3 scintillation detectors are brought into discriminator motherboard and ultra fast discriminator is used separately for RPC detector. The shaped pulses from the discriminators are fed to the fast coincidence circuit. Stretched pulses at the output of the coincidence unit corresponding to the trigger and RPC coincident with trigger are counted separately. The high voltage to the scintillator is supplied from ECIL, HV 4800D HV power supply and the RPC is biased from the HV power supply developed by us using fly back converter topology.

The variation of efficiency with applied high voltage for different ratios of gas mixture is shown in Fig. 8.3. The muon trigger rate is low; counting is performed with counter timer (ECIL 5104). For counting the RPC coincident pulses the DSO with GPIB interface is used. Whenever a trigger occurs at the scope for RPC coincident pulse the oscilloscope increments a counter set up to record the trigger. This action is continued automatically for a predetermined period. Separate software is developed for the scope to handle data acquisition system.

8.4 Conclusion

We have successfully built and characterized the glass RPCs of 15 cm × 15 cm in area. We have calculated efficiency of RPCs at different values of the high voltage. It is observed that the Parameters which have been varied in efficiency test are Gas mixture and detector bias voltage. The value of efficiency increases with respect to the high voltage and values are almost constant at 7.5 kV which is more than 92%.

Acknowledgements Special acknowledgement is due to Prof. K Boruah, and Dr. S Rajbongshi for their encouragement and support.

References

1. B. Bilki, F. Corriveau, B. Freaud et al., Tests of a novel design of resistive plate chambers. JINST **10**, P10037 (2015)
2. D. Krasnopevtsev, Studies of the gas volume current density in the ATLAS Resistive Plate Chamber detector during the 2018 data—taking at the LHC. JINST **15**, C10024 (2020)
3. P. Fonte, Application and new developments in resistive plate chambers. IEEE Trans. Nuclear Sci. **49**(3), 881–887 (2002)
4. B. Satyanarayana, INO: a national mega science and engineering project. IEEE India Info. **13**(3), 32–37 (2018)
5. A. Artamonov et al., The time response of glass resistive plate chambers to heavily ionizing particles. JINST **2**, P10004 (2007)
6. S.H. Thoker, B. Satyanarayana et al., Characteristics and performance studies of Glass resistive plate chambers. Proc. DAE-BRNS Sympos. Nucl. Phys. **60**, 1040–1041(2015)
7. H.Ch. Medhi, S.C. Rajbongshi, design and development of high voltage power supply and its application in RPC. IJAREEIE **9**(8), 2197–2201 (2020)

Chapter 9
Incorporation of Rubidium in the Organic–Inorganic $FAPbI_3$ Structure for Stabilizing the Optically Active Perovskite Phase

Ujjal Das and Asim Roy

Abstract Organic–inorganic halide perovskite (OIHP) materials have become the focus of attention in the photovoltaic and other electronic/optoelectronic applications since past few years. The exceptional and fundamental properties of these materials pave the desired state-of-the-art in the perovskite based applications. However, the associated issue of stability must be solved for commercialization. The critical weakness of the OIHPs against moisture originates mainly from the hygroscopic organic cations in the structure. In this work, rubidium (Rb) has been partially substituted at the A-site of formamidinium lead iodide ($FAPbI_3$) structure and found to be suitable for moisture-tolerance. The incorporation of Rb, induced the absorption over the entire visible window, which may lead to various long-term applicability in solar cells, photocatalysis and other optoelectronic devices.

9.1 Introduction

Last few years have perceived an unprecedented advancement in the photovoltaic arena using halide perovskites (HPs), as the primary semiconductor of interest due to thereof facile processing, tunable bandgap and superior charge-transfer properties [1–3]. In particular, the organic–inorganic HP (OIHP) materials have established its significance in research community and assure reliability in various applications [4, 5]. Despite of the superior performance, OIHP materials cannot resist the ambient temperature and moisture [6]. Several experiments have successfully explained the origin of the instability in OIHPs [7]. The key reason for the instability is believed to be due to the hygroscopic nature of the organic cation [7]. In this context, various methods have also been implemented to sustain the perovskite structure, viz. passivation, encapsulation etc. [8, 9]. But, these methods might not provide the solution in long-term usage. However, the partial substitution of the organic cation with an inorganic one resulted enhanced stability in harsh environment. Saliba et al. recently

U. Das · A. Roy (✉)
Department of Physics, National Institute of Technology, Silchar 788010, Assam, India

R. G. Nair et al. (eds.), *Proceedings of 28th National Conference on Condensed Matter Physics*, Springer Proceedings in Physics 269,
https://doi.org/10.1007/978-981-16-5407-7_9

reported that incorporation Rb at the A-cationic site of an OIHP has improved the photovoltaic performance of the perovskite solar cell [10]. In this work, the oxidation-stable rubidium (Rb^+) has been embedded at the A-site of the $FAPbI_3$, since pure $FAPbI_3$ decomposes in high temperature and humid condition.

9.2 Experimental Details

(i) Preparation of FAI—3 g of FA-acetate salt was mixed with 7 ml of HI and stirred for 10 min at 50 °C. The solution was then dried in vacuum at 100 °C for 2 h and cleaned with diethyl ether. The resultant product was again dried and kept in vacuum.

(ii) Preparation of $FAPbI_3$—A mixture of 0.2 mmol FAI, 0.2 mmol PbI_2, 40 μL n-octylamine and 0.75 mL oleic acid was dissolved in 5 mL of DMF, sonicating and forming a uniform precursor solution. Then, 50 μL precursor solution was gradually dropped into 5 mL toluene with vigorous stirring. Immediately, the yellowish-green precursor solution became red in color. But, after few minutes the red solution became transparent, showing rapid degradation at 70% humidity as shown in Fig. 9.1.

So, the next aim was to stabilize the perovskite structure of $FAPbI_3$. Therefore, 20% Rb was doped in the $FAPbI_3$.

(iii) Preparation of $FA_{1-x}Rb_xPbI_3$—FAI, RbI and PbI_2 were mixed in appropriate ratio in 5 mL DMF and 40 μL n-octylamine and 0.75 mL oleic acid were added. Then, 50 μL precursor solution was gradually dropped into 5 mL toluene with vigorous stirring. Immediately, the yellowish-green precursor solution became yellowish-red in color. This time the color of the solution didn't change after few minutes, and the precipitate were retrieved successfully as shown in Fig. 9.2.

Fig. 9.1 Experimental demonstration of $FAPbI_3$ synthesis

Fig. 9.2 Experimental demonstration of $FA_{0.8}$ $Rb_{0.2}PbI_3$ synthesis

9.3 Results and Discussion

The optical characteristics of the pure and Rb doped $FAPbI_3$ is shown in Figs. 9.3 and 9.4. The UV–visible spectroscopy was carried out to investigate the light absorbing

Fig. 9.3 UV–vis absorption spectrum of the pure and Rb doped $FAPbI_3$

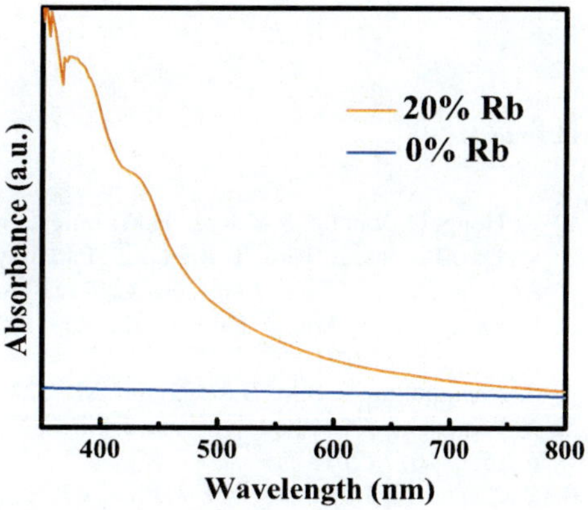

Fig. 9.4 Photoluminescence spectrum of the $FA_{0.8}$ $Rb_{0.2}PbI_3$

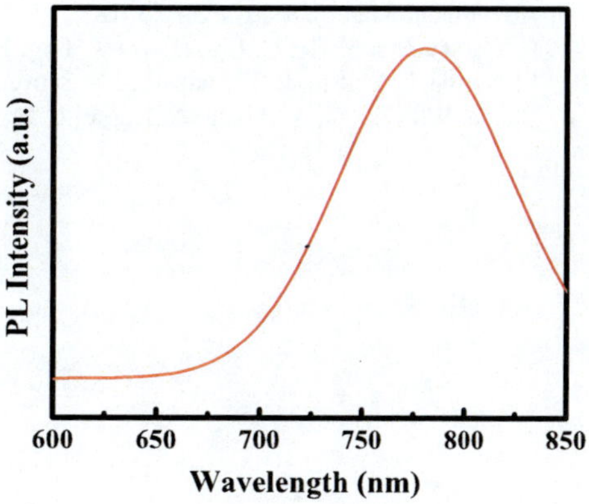

feature of the perovskite material. Figure 9.3 shows the UV–visible absorption spectrum of the 0% and 20% Rb doped $FAPbI_3$ perovskite. From figure we observe that the solution having pure $FAPbI_3$ did not show significant absorption, which may be due to the decomposition of the perovskite. However, significant band edge absorption has been observed for the $FA_{0.8}Rb_{0.2}PbI_3$ over the entire visible range. Figure 9.4 shows the photoluminescence spectrum of the $FA_{0.8}Rb_{0.2}PbI_3$ perovskite, having an emission peak at ~780.48 nm.

9.4 Conclusion

In summary, we have synthesized the organic–inorganic halide perovskite (OIHP) $FAPbI_3$ and found to be unstable at ambient environment. On the other hand, the Rb doped $FAPbI_3$ didn't decompose in the ambient atmosphere and found to be optically active in the visible region. Thus, the Rb doped $FAPbI_3$ possesses the potential application in various electronic and optoelectronics.

References

1. K. Hong, Q. Van Le, S.Y. Kim, H.W. Jang, J. Mater. Chem. C **6**, 2189 (2018)
2. U. Das, D. Das, B. Paul, T. Rabha, S. Pattanayak, A. Kanjilal, S. Bhattacharjee, P. Sarkar, A. Roy, ACS Appl. Mater. Interfaces **12**, 41718–41727 (2020)
3. U. Das, A. Nyayban, B. Paul, A. Barman, P. Sarkar, A. Roy, ACS Appl. Electron. Mater. **2**, 1343–1351 (2020)
4. S. Ananthakumar, S.M. Babu, Synth. Met. **246**, 64–95 (2018)
5. Y. Wang, Z. Lv, L. Zhou, X. Chen, J. Chen, Y. Zhou, V.A.L. Roy, S.T. Han, J. Mater. Chem. C **6**, 1600–1617 (2018)
6. J. Bisquert, E.J. Juarez-Perez, J. Phys. Chem. Lett. **10**(19), 5889–5891 (2019)
7. A. Ciccioli, A. Latini, J. Phys. Chem. Lett. **9**(13), 3756–3765 (2018)
8. J.S. Han, Q.V. Le, J. Choi, K. Hong, C.W. Moon, T.L. Kim, H. Kim, S.Y. Kim, H.W. Jang, Adv. Funct. Mater. **28**, 1705783 (2018)
9. H. Cai, G. Ma, Y. He, C. Liu, H. Wang, Org. Electron. **58**, 301–305 (2018)
10. M. Saliba, T. Matsui, K. Domanski, J.Y. Seo, A. Ummadisingu, S.M. Zakeeruddin, J.P. Correa-Baena, W.R. Tress, A. Abate, A. Hagfeldt, M. Grätzel, Science **354**, 206–209 (2016)

Chapter 10
Structure, Interband Transition Strength and Estimation of ELF of Perovskite Thin Film of $CCT_{1\text{-}x}Nb_xO$ for x = 0.02

D. Dwibedy, A. K. Sahoo, and Manas R. Panigrahi

Abstract Niobium modified calcium copper titanate ($CCT_{1\text{-}x}Nb_xO$) thin film was prepared by a non-conventional sol–gel Route (NCSGR) and the material was observed to be crystallized in a crystalline perovskite structure. The diffraction pattern of the prepared thin film was subjected to Rietveld refinement for quantitative structural analysis. The optical properties of the synthesized material were studied using the Uv–Vis spectrophotometer. Different optical properties of the studied material were analysed but here, the study of inert-band transition strength and energy loss function was focused and explained here.

10.1 Introduction

Significant attention has been given to ideal material and their dielectric isomorphic properties and significance for the miniaturisation of capacitor, including other electronic devices. Recently, researcher and scientist have been trying to develop material which can harvest more energy for solar cell application. In this process, a good number of materials have been developed for efficient absorption of ultraviolet rays. A $Cu_3Ti_4O_{12}$ oxide compound has become a valuable candidate that can provide a considerably higher dielectric constant, showing a complex perovskite structure and thus a flexible candidate for various possible applications. Several researchers have currently documented various methods for the manufacturing of the material due to its huge dielectric factors [1–4]. The traditional method of the solid state reaction from these techniques has several demerits in order to preserve the homogeneity of the starting material. For the solid state diffusion to take place, a single phase requires more time to sinter as well as higher sintering temperature. Consequently, the right chemical method with lower processing temperatures and processing times is required for better homogeneity, which can be both time and

D. Dwibedy · A. K. Sahoo · M. R. Panigrahi (✉)
Department of Physics, Veer Surendra Sai University of Technology, Burla, Sambalpur 768018, India
e-mail: mrpanigrahi@vssut.ac.in

R. G. Nair et al. (eds.), *Proceedings of 28th National Conference on Condensed Matter Physics*, Springer Proceedings in Physics 269,
https://doi.org/10.1007/978-981-16-5407-7_10

energy saving. Since researchers doubt that when these materials are used at higher concentration the UV rays get absorbed in the body. Hence research attentions have been focused to develop harmless UV absorbents. Since some nanoparticles such as TiO_2, ZnO and Ce_2O have the ability to screen off UV rays (wavelength < 400 nm) effectively, their nanomaterial have been commercially used in cosmetic products. ZnO and TiO_2 generate reactive oxygen species highly because of their efficient photocatalytic nature. Furthermore, the high RI (refractive index) of TiO_2 have the ability to turn the skin naturally white. Likewise, the material Cao modified CeO_2 is transparent to the visible light having band gap energy 3 eV (approx.) and less photocatalytic properties. This makes the material to be used as a UV blocking material commercially [5]. Here, in this paper it is attempted to prepare a single phase $CaCu_3Ti_4O_{12}$ (CCTO) and Nb modified CCTO with formula $CCT_{1-x}Nb_x$ for $x = 0.02$ ceramic thin film for photovoltaic application as it is an energy harvesting material with interesting dielectric properties.

10.2 Experimental

Nb-doped CCTO (Nb_xCCTO_{1-x} for $x = 0.02$) thin film was synthesized by the modified sol–gel method. The interacting reagent materials were $CaCO_3$ (Merck, 99.99%), Cu_2O (Merck, 99.99%), TiO_2 (Merck, 99.99%) and Nb_2O_5 (Merck, 99.99%). Initially, the reagents materials were blended in a stoichiometric ratio and the acetic acid was applied in certain amounts(~5 ml.) to achieve an even and lump-free paste. Deionized (DI) water was then applied dropwise to the paste with continuously stirred at room temperature by magnetic stirrer, until the clear, uniform sol was obtained. Then, the sol was added with 0.1 M HNO3 and subjected to 6–7 h at 180 °C for refluxing. The Nb-doped CCTO material in gel form, was eventually prepared and after being kept for one day, the synthesized material was used to coat the ITO glass slides using the doctor's blade technique to get the desired Nb-doped CCTO thin film. Then, the 200 °C annealed thin film of the synthesized Nb-CCTO was subjected to XRD (Shimadzu-6100) with CuK_α radiation ($\lambda = 0.1540$ nm) to analyse the crystal structure. For optical characterization of the prepared thin film, the UV–Vis spectrometer (UV3092) was used within 300–850 nm wavelength range with the 5 nm of spectral bandwidth. however, the performed scanning interval was of 2 nm with medium speed.

10.3 Results and Discussion

The x-ray diffractogram of $CaCu_3Ti_4O_{12}$ (CCTO) and Nb doped $CaCu_3Ti_4O_{12}$ (Nb-CCTO) thin films is shown in Fig. 10.1a. It is observed that the material crystallizes in cubic perovskite structure in space group (H-M): Im-3 m, Hall: P1, Sch: C1^1 and cubic crystal system. Further, the lattice parameters of the synthesized material

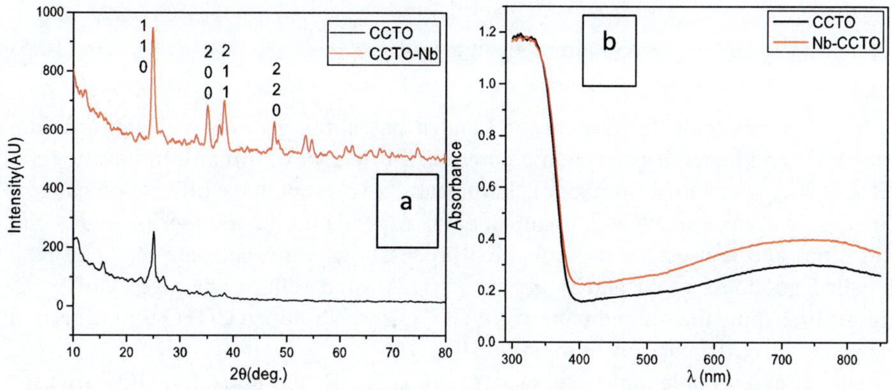

Fig. 10.1 X-ray diffractogram (**a**) and the absorbance spectra (**b**) of CCTO and Nb doped CCTO thin film

were found to be a = b = c = 5.2418 A°; α = β = γ = 90°, with unit cell volume = 144.0261A°³. The crystallite size of the material was estimated to be 101.7 nm using the well-known Debye Scherrer equation [6, 7].

$$t_{hkl} = \frac{K\lambda}{\beta \mathrm{Cos}\theta_{hkl}} \tag{10.1}$$

For the most intense peak, the r.m.s. strain is estimated to be 1.32 × 10^{-3}. The non-linear least square method has been adopted for Rietveld refinement and for convergence solution of multiple iterations for refinement is considered. A numerical value that explains the best fit between the observed and calculated pattern during the refinement process is quantified and is called the figure of merit. Few terms generally used for the representation of figure of merit are given below which indicate the quality of refinement, or in other words, these values show the degree of fit between the observed and calculated data.

The different Profile Residual (reliability factor) are explained with equations as follows[8];

$$\text{Weighted profile residual}: \; w_{ss} = \sum\nolimits_{(i=1)}^{N} [w_i(I_i^{exp} - I_i^{calc})]^2, \qquad w_i = \frac{1}{\sqrt{I_i^{exp}}} \tag{10.2}$$

$$\text{Bragg residual}: \; R_{wp} = \sqrt{\frac{\sum_{i=1}^{N}\left[w_i\left(I_i^{exp} - I_i^{calc}\right)\right]^2}{\sum_{i=1}^{N}\left[w_i I_i^{exp}\right]^2}} \tag{10.3}$$

$$\text{Expected profile residual}: \; R_{exp} = \sqrt{\frac{N - P}{\sum_{i=1}^{N}\left[w_i I_i^{exp}\right]^2}} \tag{10.4}$$

$$\text{Goodness of fit : GoF} = \frac{R_{wp}}{R_{exp}} \tag{10.5}$$

R_{wp} depends upon the background and it has nothing to do with the absolute intensities and hence it is a reliable parameter. It is always difficult to have a good value of R_{wp} when more number of sharp peaks are present in the diffraction pattern. The absolute value of weighted sum squares depends on the number of points and intensities and is used for minimization process. The ratio between R_{wp} and R_{exp} is called goodness of fit and is always ≥ 1. A good refinement gives GoF ≤ 2. Figure 10.2 show the refined patterns of CCTO and Nb-doped CCTO thin films and the quantitative data are given in Table 10.1.

The refined reliable parameters such as R_{wp}, R_p, R_{wpnb}, R_{pnb}, and GOF are estimated using Maud 2.92 and presented in Table 10.1 above, where the symbols have their usual meaning.

Figure 10.1b represents the absorption spectra of both CCTO and Nb-CCTO. Figure 10.3 represents the real and imaginary part of inter-band transition strength CCTO and Nb-doped CCTO thin films. The inter-band transition strength is important to be accounted for the rule of dipole selection for the transitions and also related to the possibility of an electron transition among the conquered valance band and the unoccupied conductive band with transition energy. The considered optical transition for inter-band transitions is of two types and occurs among the conduction band

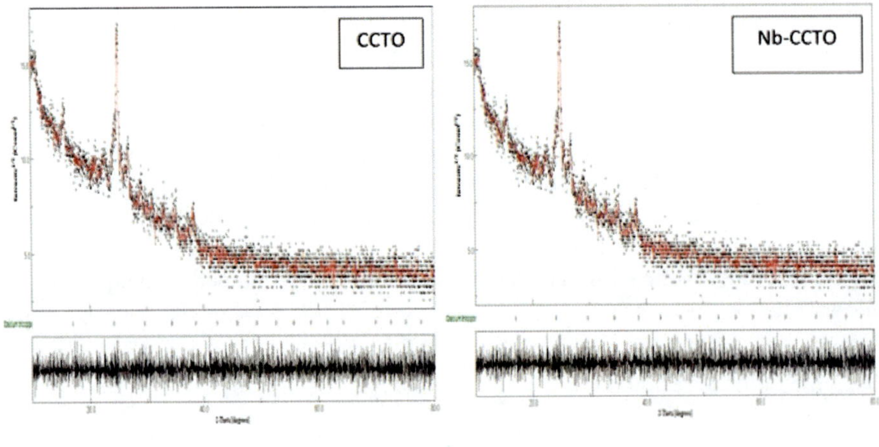

Fig. 10.2 Refined diffractogram of CCTO and Nb doped CCTO thin films

Table 10.1 Structural parameters of the thin films of CCTO and Nb doped CCTO

Samples	2θ (deg.)	β	t_{hkl} (nm)	Strain	R_{wp}	R_p	R_{pnb}	R_{wpnb}	Gof
CCTO	25.14	0.144	53.54	2.82	0.293	0.223	0.759	0.589	1.42
CCTO-Nb	24.73	1.152	3.59	22.93	0.293	0.238	0.996	0.998	1.65

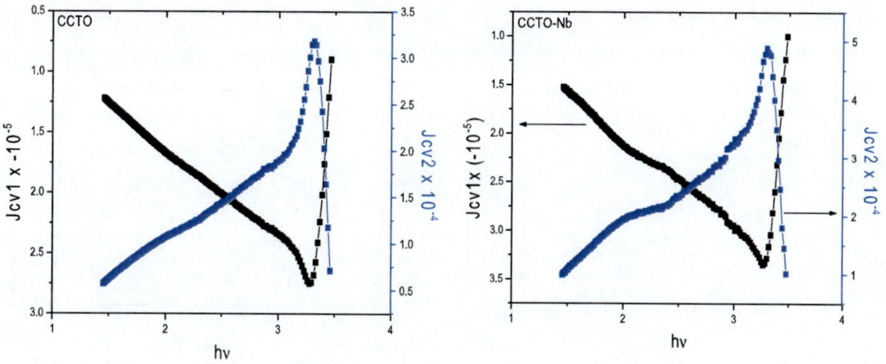

Fig. 10.3 Inter-band transition strength of CCTO and Nb doped CCTO thin films

and the valance band involving both the carriers, i.e. holes and electrons and hence transition is bipolar. Either within the conduction or within the valance band, the inter-band transitions include just one kind of carrier and so the transition is unipolar. The excitation of electrons occurs due to the interaction of the photon with the material. These excited electrons jump into unoccupied energy levels in the conduction band including the collective excitation of valance electrons. The interband transition is initiated and identified as a transition in a band structure model, from the excitation of electrons in the valance band to a vacant state in the conduction band. The inter-band transition strength of the material represents the optical response of the material and it displays the valence to conduction band transition of the inert-band electronic structure. The inter-band transition strength (J_{cv}) is dependent on the dielectric constant as [9]

$$J_{cv}(E) = \frac{(m_0)^2(E^2)}{(e^2)\left(\frac{h}{2\pi}\right)^2(8\pi^2)}(\varepsilon_2 + i\varepsilon_1) \tag{10.6}$$

The inter-band transition strength is proportional to the transition probability. Factor $m_0^2 e^{-2}\left(\frac{h}{2\pi}\right)^{-2}$ whose value is 8.289×10^{-6} in CGS unit is taken as unity for ease of calculation. The inter-band transition strength (Re) initially decreases with increasing photon energy and increases rapidly at 2.5 eV and reflects the probability of more number of electron transitions at photon energy more than 2.5 eV. Thereby the absorption becomes more after this particular value of energy and excitations of electrons from valence band to conduction band increases. Surface energy loss function (SELF) and the volume energy loss function (VELF) is also estimated using the ε_1 and ε_2 [10], as given below in (10.7) and (10.8),

$$\mathrm{SELF} = -\mathrm{Im}\left(\frac{1}{1+\varepsilon}\right) = \frac{\varepsilon_2}{((\varepsilon_1 + 1)^2 + \varepsilon_2^2)} \tag{10.7}$$

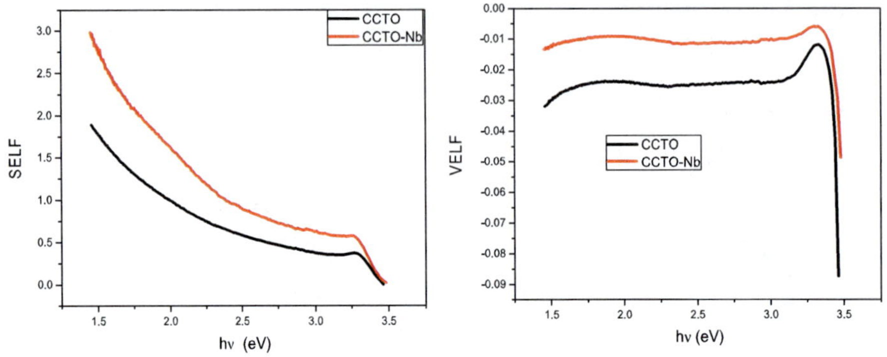

Fig. 10.4 Surface and volume energy loss function of CCTO and Nb doped CCTO thin films

$$\mathrm{VELF} = -\mathrm{Im}\left(\frac{1}{\varepsilon}\right) = \frac{\varepsilon_2}{(\varepsilon_1 + \varepsilon_2)^2} \tag{10.8}$$

There are two regions of electronics energy loss function. The high energy loss region (>1 eV) is the first region, and the low energy loss region (<1 eV) is the second region. The first region study after ionization edge can provide important and clear information about the oxidation state and the analysis of the second region can provide data on the electronic structure and also composition. Generally, a material's energy loss feature arises from the dielectric constant of the material and within the context of dielectric concept validity, an inelastic scattering mechanism is experienced, which is characterised by a transfer of momentum. The VELF and SELF always show the increased value with increased photon energy, but it is seen that after a particular energy VELF decreases sharply so also the SELF. However, the VELF values are always greater than the SELF values. The behaviour of SELF and VELF with respect to photon energy illustrates the energy at which maximum extinction occurs and this is 3.28 eV for SELF and 3.34 eV for VELF. Both the bulk and the surface of the material lose energy when the electrons move from the valence band to the conduction band at 3.28 eV and 3.34 eV. The energy losses owe to the crossover of the first valance electron to the conduction band in the material of the films. The various optical parameters discussed above gives an idea about the potential application of the synthesized material for energy harvesting from the solar cell (Fig. 10.4).

10.4 Conclusion

In this paper, the studied material $(CaCu_3Ti_4O_{12})_{1-x}\,Nb_x$ for $x = 0.02$ is synthesized with a novel technique with ease of fabrication, low cost, time-saving and less energy consumption to a high purity single phase material. The structure of the material is observed to perovskite and the various structural data is presented in this paper.

Various optical parameters are studied and keeping the page constraints in view only two parameters such as inter-band transition strength and energy loss function are presented. VELF and SELF show a singular behaviour with increasing photon energy up to a particular value of 3.34 eV and 3.28 eV which shows the extinction energy. So, the bulk and the surface of the material lose energy when the electrons move from the valence band to the conduction band at the mentioned energies. The energy losses owe to the crossover of the first valance electron to the conduction band in the material of the films. The various optical parameters discussed above gives an idea about the potential application of the synthesized material for energy harvesting from the solar cell.

References

1. A.K. Rai, K.D. Mandal, D. Kumar, O. Parkash, J. Alloys Compd. **491**, 507 (2010)
2. A. Sen, U.N. Maiti, R. Thapa, K.K. Chattopadhyay, J. Alloys Compd. **506**, 853 (2010)
3. A.K. Mishra, D. Dwibedy, M. Devi, M.R. Panigrahi, Trans. Electr. Electron. Mater. **21**, 315 (2020)
4. A. Moquim, M.R. Panigrahi, J. Mater. Sci. Mater. Electron. **26**, 4956 (2015)
5. S. Yabe, T. Sato, J. Solid State Chem. **171**, 7 (2003)
6. M. Devi and M. R. Panigrahi, IOP Conf. Ser. Mater. Sci. Eng. **653** (2019).
7. M. Devi, M.R. Panigrahi, U.P. Singh, J. Mater. Sci. Mater. Electron. **26**, 1186 (2014)
8. M.R. Panigrahi, S. Panigrahi, Phys. B Condens. Matter **405**, 3986 (2010)
9. M.A. Nattiq, J. Basrah Res. **40**, 28 (2014)
10. M. F. Al-mudhaffer, J. Basrah Res. **36** (2010).

Chapter 11
Investigation of Proton Transport Mechanism in Protic Ionic Liquid Based Polymeric Nanocomposite Membranes

Manjula G. Nair, Rajesh Deb, Tohru Tsuruoka, and Saumya R. Mohapatra

Abstract Proton conducting polymer electrolytes play a key role in the development of new generation supercapacitors and proton exchange membrane fuel cells (PEMFCs). For the development of better electrolyte membrane, understanding of proton transport mechanism in them is very important. In this study, we have developed proton conducting non-aqueous electrolytes based on protic ionic liquid diethylmethylammoniumtrifluoromethanesulfonate ([dema][TfO]), montmorillonite clay (MMT), phosphoric acid and poly (vinylidene fluoride-co-hexafluoropropylene); PVDF-HFP as the host polymer. The proton transport mechanism is investigated from the temperature and frequency dependence of electrical conductivity data. The ac conductivity spectra follow Jonscher power law. With suitable scaling, the ac conductivity spectra show single master curve, which agrees to Time–Temperature superposition principle. The dc ionic conductivity measured at 20 °C in vacuum for 1 wt. % of MMT incorporated membrane is found to be 3.05×10^{-5} S cm^{-1}. This is five times higher than that of 0.5 wt. % MMT incorporated membrane (6.0×10^{-6} S cm^{-1}). The activation energies for dc conductivity are estimated to be 31.4 kJ/mole and 21.8 kJ/mole for membranes with 0.5 and 1.0 wt. % MMT, respectively. This suggests that the proton transport process may be dominated by Grotthuss mechanism.

11.1 Introduction

Sustainable, renewable, eco-friendly fuels and energy sources are now the global need to address the pollution and energy requirements [1, 2]. There is an increased research attention recently on protic ionic liquids (PILs) because of their versatile

M. G. Nair · R. Deb · S. R. Mohapatra (✉)
Department of Physics, National Institute of Technology Silchar, Silchar 788010, Assam, India
e-mail: saumya@phy.nits.ac.in

T. Tsuruoka
International Center for Materials Nanoarchitectonics (WPI-MANA), National Institute for Materials Science, 1-1 Namiki, Tsukuba 305-0044, Japan

R. G. Nair et al. (eds.), *Proceedings of 28th National Conference on Condensed Matter Physics*, Springer Proceedings in Physics 269,
https://doi.org/10.1007/978-981-16-5407-7_11

use in the area of energy applications. PILs incorporated polymer electrolytes are ideal electrolytes in the proton exchange membrane fuel cells (PEMFCs) because of their high proton conductivity even in the anhydrous conditions and excellent electrochemical and thermal properties. The blending of PILs with different polymers have been studied to obtain the optimal properties. Compared to widely used perfluorosulfonicacid (PFSA) based electrolytes in PEMFCs, it may be possible to increase the operating temperature range of PEMFCs to above 100 °C and improve the efficiency with these electrolytes. In order to address the leaching of PILs in the electrolytes, addition of different fillers have been reported. This helps to withhold more amount of ionic liquids in the membranes as well as improves the mechanical and thermal stabilities of the electrolytes [3–5].

Understanding fundamental aspects of proton transport mechanism in PEM holds paramount importance for better material development. The proton conduction mechanism studies are mostly carried out at microscopic level with quasi-elastic neutron scattering (QENS) and NMR investigations. However, temperature dependent electrical conductivity study can also be effective to understand the transport properties, although it is an indirect way. The proton transport in polymer electrolytes can be described based on three main mechanisms: (a) proton hopping or Grotthuss mechanism (b) diffusion or vehicular mechanism, and (c) direct transport via polymer chain segmental motions. In the present study, we prepared proton conducting nonaqueous polymer composite electrolytes based on protic ionic liquid [dema][TfO], montmorillonite clay (MMT), phosphoric acid (H_3PO_4), and PVDF-HFP as the host polymer. The proton transport mechanism was investigated from the temperature dependent dc ionic conductivity studies. Previous literatures suggested that the proton transport mechanism could be Grotthuss type when the activation energy of dc ionic conductivity falls in the range of 14–40 kJ/mole [6–8]. In some polymer electrolytes, proton transportation occurs via all three above mentioned mechanisms [8].

11.2 Materials and Methods

The polymer poly (vinylidene fluoride-co-hexafluoropropylene); PVDF-HFP (MW; 400,000), Phosphoric acid; H_3PO_4 (neat; 99.99%) and montmorillonite clay; MMT K10 (surface area 220–270 m^2/g) were procured from Sigma-Aldrich. The protic ionic-liquid diethylmethylammoniumtrifluoromethanesulfonate; [dema][TfO] was supplied by IoLiTec and the solvent acetone was obtained from Merck. The polymer electrolyte containing PVDF-HFP, H_3PO_4, and [dema][TfO] is the primary electrolyte (PE) system. Series of polymer electrolytes were prepared by addition of different concentrations of MMT into the PE system. Sufficient amount of PVDF-HFP with acetone was taken in a flask with a screw cap. It was kept for magnetic stirring until the polymer is completely dissolved. Further 10 wt. % of the neat H_3PO_4 was added to the solution using a micropipette and kept for magnetic stirring. Then, 70 wt. % of PIL; [dema][TfO] was added in to the flask and magnetically stirred to obtain a homogeneous mixture. Different concentration of the filler MMT (0.5

and 1 wt. %) was added in the PE solution and continued the stirring to obtain a dispersed solution. Finally, free standing polymer electrolyte membranes (PEMs) were obtained by solution casting technique followed by drying them in a vacuum.

Vibrational properties of the prepared samples were investigated by Fourier-transform infrared (FTIR) spectroscopy in the range of 400–4000 cm^{-1} using Jasco 4700 instrument. Temperature dependent complex impedance analysis was performed in a frequency range of 1 Hz–10^7 Hz using a home-made cell equipped with Novocontrol Alpha-A analyzer.

11.3 Results and Discussions

11.3.1 FTIR Analysis

To investigate the interaction between PVDF-HFP, H_3PO_4, [dema][TfO], and MMT at the molecular level, FTIR spectroscopy was employed. The infrared spectra of the prepared electrolyte membranes in the range of 500–1500 cm^{-1} and 2500–3500 cm^{-1} are shown in Fig. 11.1a, b, respectively. The FTIR spectra are quite sensitive to different polymorphs of the host polymer, due to their different chain conformation. In the range of 500–1500 cm^{-1}, the characteristic peaks corresponding to α (TGT$\bar{G}$), β (TTTT), and γ ($T_3GT_3\bar{G}$) are very much useful for phase identification and their quantification. However, as some of the absorption peaks are commonly observed for all the three phases or at least for two of them, focus must be to find undisputed exclusive peaks for individual phases [9–11]. The presence of PIL in the electrolyte membranes is manifested in terms of appearance of 575 cm^{-1} and 639 cm^{-1} peaks, owing to asymmetric deformation of CF_3 and symmetric deformation of SO_3 present

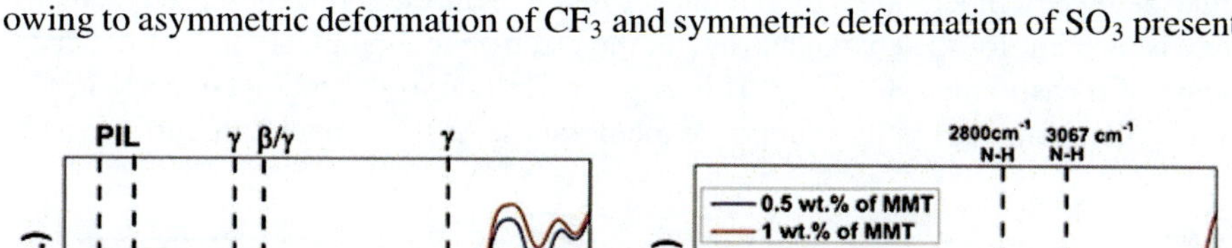

Fig. 11.1 FTIR spectra of polymer composite electrolytes in the range **a** 500–1500 cm^{-1} and **b** 2500–3500 cm^{-1}

in the [dema][TfO], respectively [12]. The peak at 833 cm^{-1} is assigned to γ phase of PVDF-HFP. The peak due to both β and γ phases of the polymer are observed at 875 cm^{-1} [10]. Another characteristic band of the PIL is due to N–H stretching observed at 2800 cm^{-1} and 3067 cm^{-1} [Fig. 11.1b]. A broad peak due to residual water is observed at ~3500 cm^{-1}. The exclusive peak for the γ phase is observed in all the PEM films at ~1230 cm^{-1}. The peaks observed at 1077 cm^{-1} and 1164 cm^{-1} are due to the CF3 out of plane deformation and C-F stretching, respectively [13, 14]. The exclusive α phase peak that usually observed around ~760 cm^{-1} in PVDF-HFP is absent in both the polymer composite membranes. FTIR analyses indicate that the host polymer in the PEM preferably crystallizes into polar β and γ phases than non-polar α phase [15]. From the FTIR studies it is found that there is no significant difference in the spectra observed between the two MMT concentrations.

11.3.2 Electrical Conductivity Studies

Complex impedance spectra (Z' vs. Z") of proton conductive composite electrolytes over a temperature range of −60 °C to 80 °C are shown in Fig. 11.2. All the data were obtained from the samples sandwiched between blocking Pt electrode on both sides, which were placed under low vacuum conditions (<10 Pa). In both the composite membranes, at low temperature (−60 °C and −40 °C), the spectra are distinctly composed of a single semicircle at higher frequency side followed by a spike at lower frequency side which could be an arc of bigger semicircle. With temperature rise, the semicircle in the high frequency side vanishes. This also marked by the shifting of linear spike to higher frequency with increasing slope. The presence of semicircle and spike at lower temperature region indicates that two different relaxation mechanisms may be present for the charge carriers in the electrolyte membrane in the frequency range of measurement [16, 17]. This is probably due to the inhomogeneity in the conduction pathways in the composite membranes. As the temperature further rises,

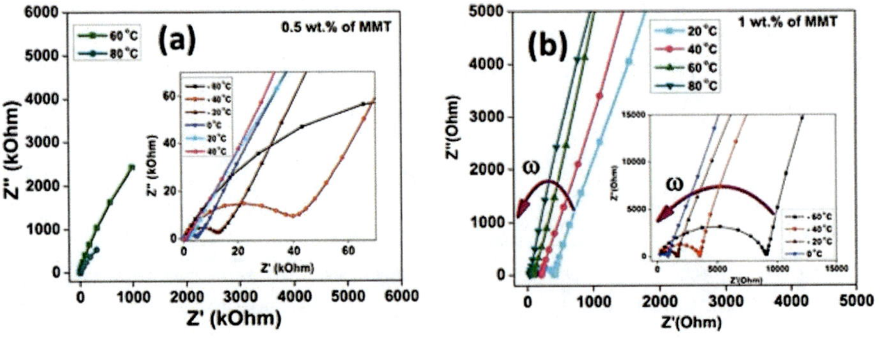

Fig. 11.2 Complex impedance spectra of polymer composite electrolytes with **a** 0.5 wt. % of MMT and **b** 1 wt. % of MMT for various temperature

only one kind of relaxation mechanism is likely to present due to improved polymer flexibility. The increase in slope of the spikes suggests that the composite membranes behave more like a capacitive device than resistive one. The increase in temperature makes the polymeric matrix more flexible and support faster charge transport. The dc ionic conductivity is calculated from the data and found to be increasing with the temperature. The dc ionic conductivity measured at 20 °C for 1 wt. % and 0.5 wt. % of MMT are evaluated to be ~3.05×10^{-5} S cm^{-1} and ~6.0×10^{-6} S cm^{-1}, respectively.

The relaxation mechanism in the composite membranes can be better investigated using loss-tangent spectra, as shown in the Fig. 11.3. In both the composite membranes, the loss tangent peak observed to shift to higher frequency with increased temperature. The loss tangent peak originates from the conductivity relaxation and is thermally activated. This suggests that the underlying relaxation mechanism remains the same even if we change the temperature. Such behavior is known as Time–Temperature superposition principle (TTSP) and; the loss tangent and AC conductivity spectra may follow some scaling laws [18, 19].

The AC conductivity spectra of the composite membranes are presented in the Fig. 11.4. The spectra show typical Jonscher power law behavior at higher frequency region. The hopping frequency, dc conductivity, and power law exponent (n) are calculated by fitting the ac conductivity spectra with Jonscher relation:

$$\sigma_{ac}(\omega) = \sigma_{dc} + A\omega^{n},$$

where, $A\omega^n$ is pure dispersive component of AC conductivity, A is a constant which determines the strength of polarizability, n is a power exponent ($0 \leq n \leq 1$) describing degree of interaction between mobile ions and the polymer. The low frequency dispersion observed in the spectra are due to space charge accumulation at the electrode interface. The frequency range over which this dispersion observed increases with increasing temperature. To verify the TTSP, we plotted the master curves for ac

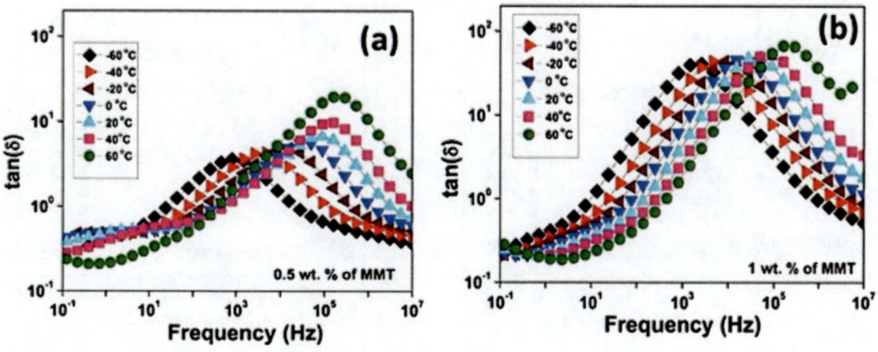

Fig. 11.3 Loss tangent spectra (tan δ) of **a** 0.5 wt. % MMT and **b** 1 wt. % MMT added polymer composite electrolytes at various temperatures

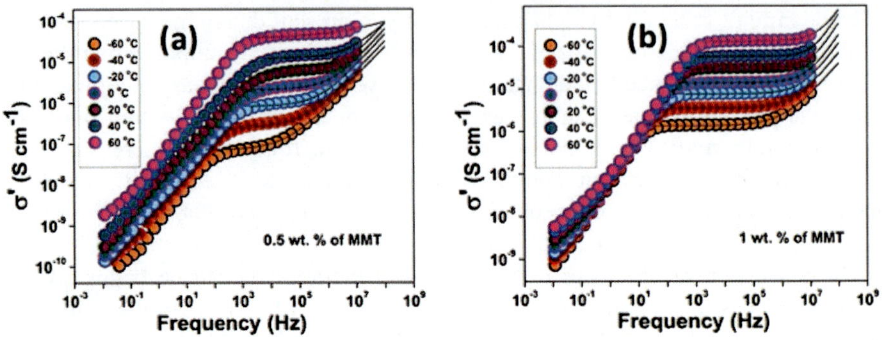

Fig. 11.4 Frequency dependence of the ac conductivity spectra of **a** 0.5 wt. % and **b** 1 wt. % MMT added polymer composite electrolytes at different temperatures. The solid grey lines are fits of the experimental data to JPL equation

conductivity by suitably scaling the ac conductivity and frequency axis, as shown in Fig. 11.5. It is found that a single master curve is observed in the high-frequency region (where Jonscher power law is obeyed) over the temperature range examined for both the samples. This suggests that the basic mechanism of charge transport is temperature independent. However, the low frequency dispersion does not follow a single master curve especially for 0.5 wt. % of MMT containing membrane. The low frequency dispersion is due to electrode polarization effect [20]. This can be affected by many parameters such as cell length, interfacial roughness etc., which are not intrinsic to the electrolyte [21]. That's why it is difficult to correlate the variation of low frequency ac conductivity spectra from the master curve with the MMT concentration.

To further investigate the proton or charge transport mechanism in the composite electrolyte membranes, the temperature dependence of dc conductivity and hopping frequency is investigated. The results are presented in Fig. 11.6. For both the

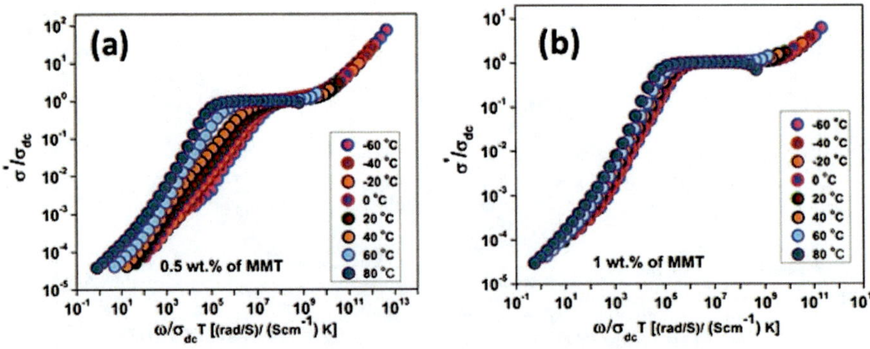

Fig. 11.5 Scaling of the conductivity spectra for **a** 0.5 wt. % MMT and **b** 1 wt. % MMT added polymer composite electrolytes at different temperatures

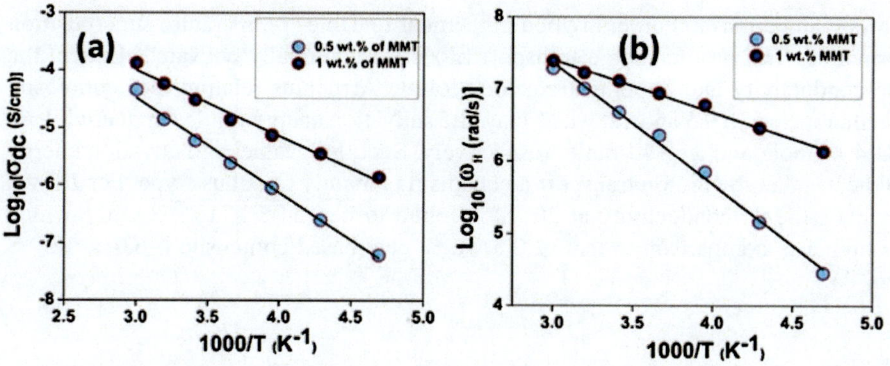

Fig. 11.6 Inverse temperature dependence of **a** dc conductivity and **b** hopping frequency for different concentrations of MMT. The solid grey lines are fits of the experimental data to Arrhenius equation

composite membranes containing 0.5 wt. % and 1 wt. % MMT, the log σ_{dc} versus *1000/T* plots are linear, which follow the Arrhenius relation:

$$\sigma = \sigma_0 exp\left(\frac{-E_a}{K_B T}\right),$$

where, σ_0 is the pre-exponential factor, E_a is the activation energy, and K_B is the Boltzmann constant. The activation energy calculated for 0.5 wt. % and 1 wt. % of MMT containing membranes are 31.4 kJ/mole and 21.8 kJ/mole, respectively. These small values of the activation energy indicates that the proton transport may be favored by Grotthuss mechanism [7, 8]. Hopping frequency also followed similar Arrhenius relation with activation energies of 33 kJ/mole and 14 kJ/mole for both the samples. For 1 wt. % of MMT based electrolyte membrane, the conductivity and hopping frequency are higher than that for 0.5 wt. % of MMT containing membrane. This suggests that the 1 wt. % MMT based electrolyte membrane possess faster charge dynamics.

11.4 Conclusions

Non-aqueous proton conductive polymeric electrolytes were prepared using PIL, H_3PO_4, PVDF-HFP with different concentrations of montmorillonite clay (MMT). The effect of clay addition is investigated using FTIR to find the interaction of clay with other constituents of the electrolyte. Presence of clay favors the polar phase of PVDF-HFP. The electrical transport properties of the composite membranes were studied from complex impedance spectroscopy data. The AC conductivity spectra follow Jonscher power law. With suitable scaling, the ac conductivity spectra shows single master curve, in good agreement to Time–Temperature superposition principle. The proton/charge transport process is thermally activated. Hence, the dc conductivity and hopping frequency follow Arrhenius relation. For composite membranes with 0.5 and 1.0 wt. % clay, the activation energy for dc conductivity are 31.4 kJ/mole and 21.8 kJ/mole, respectively. Such low values of activation energy indicates that the proton transport mechanism is mainly Grotthuss type. For 1.0 wt. % clay, the dc conductivity at 20 °C is found to be ~3.05×10^{-5} S cm^{-1} which is higher in comparison to that of 0.5 wt. % clay based composite (~6.0×10^{-6} S cm^{-1}).

References

1. S.J. Peighambardoust, S. Rowshanzamir, M. Amjadi, Int. J. Hydrog. Energy **35**, 9349–9384 (2010)
2. S.R. Narayanan, S.-P. Yen, L. Liu, S.G. Greenbaum, J. Phys. Chem. B **110**, 3942–3948 (2006)
3. H.A. Elwan, M. Mamlouk, K. Scott, J. Power Sour. **484**, 229197 (2021)
4. H. Gao, K. Lian, RSC Adv. **4**, 33091 (2014)
5. Y.P. Ying, S.K. Kamarudin, M.S. Masdar, Int. J. Hydrog. Energy **43**, 16068–16084 (2018)
6. S. Ochi, O. Kamishima, J. Mizusaki, J. Kawamura, Solid State Ion. **180**, 580–584 (2009)
7. D.-W. Lim, M. Sadakiyo, H. Kitagawa, Chem. Sci. **10**, 16–33 (2019)
8. S. Liu, L. Zhou, P. Wang, F. Zhang, S. Yu, Z. Shao, B. Yi, ACS Appl. Mater. **6**, 3195–3200 (2014)
9. L. Ruan, X. Yao, Y. Chang, L. Zhou, G. Qin, X. Zhang, Polymers **10**, 228 (2018)
10. X. Cai, T. Lei, D. Sun, L. Lin, RSC Adv. **7**, 15382–15389 (2017)
11. K. Mishra, S.A. Hashmi, D.K. Rai, J. Solid State Electrochem. **18**, 2255–2266 (2014)
12. S. Ramesh, L.J. Yi, Ionics **15**, 413–420 (2009)
13. W. Luo, L. Zhao, J. Membr. Sci **451**, 32–39 (2014)
14. A. Daneshkhah, S. Shrestha, A. Siegel, K. Varahramyan, M. Agarwal, Sensors **17**, 595 (2017)
15. S. Borah, M. Deka, Mater. Sci. Eng. B **263**, 114822 (2021)
16. M.G. Nair, S.R. Mohapatra, Mater. Lett. **251**, 148–151 (2019)
17. M.G. Nair, S.R. Mohapatra, M.R. Garda, B. Patanair, A. Saiter-Fourcin, S. Thomas, Mater. Res. Express **7**, 064005 (2020)
18. S.R. Mohapatra, A.K. Thakur, R.N.P. Choudhary, Indian J. Eng. Mater. Sci. **15**, 347–351 (2008)
19. S.R. Mohapatra, A.K. Thakur, T. Sakuma, J. Phy. Soc. Jpn. **79**(Suppl. A), 169–172 (2010)
20. P. Pal, A. Ghosh, J. Appl. Phys. **126**, 135102 (2019)
21. A.A. Khamzin, I.I. Popov, R.R. Nigmatullin, Phys. Rev. E **89**, 032303 (2014)

Chapter 12
Microwave Assisted Reduction of Graphene Oxide: Variation in Reduction Level and Enhancement in Resistive Switching Properties

Koustav Kashyap Gogoi, Nipom Sekhar Das, and Avijit Chowdhury

Abstract Resistive switching (RS) memories have emerged as a potential candidate owing to the various advantageous properties and excellent performance. Amongst the various materials exploited for RS properties, reduced graphene oxide (rGO) has gained utmost attention due to the extraordinary properties such as high mechanical strength, charge carrier mobility and so forth. Herein, we report the RS properties of rGO-PMMA nanocomposites sandwiched between top and bottom electrodes of aluminum and ITO, respectively. Each device consists of 4 cells that are fabricated in a cross-bar structure. The devices exhibit stable bipolar RS properties with excellent repeatability, low set voltage ($V_{SET} \sim -1.25$ V) and high current ON/OFF (I_{ON}/I_{OFF}) ratio of 10^4. Furthermore, the graphene oxide is reduced at three different reduction levels by varying the power of microwave irradiation (540, 720 and 900 W). The rGOs exhibit a decrease in the number of defect states as well as transition from pure amorphous to polycrystalline nature with the increase in power of microwave irradiation. The resistive switching properties also improve (V_{SET} decreased from -2 V to -1.25 V) upon increasing the level of reduction. The study highlights a cost-effective method for enhancement in memory properties through simple variation in reduction levels which may be utilized for the next generation data storage devices.

12.1 Introduction

In the recent years, organic materials have gained utmost attention in the field of RS memory owing to their advantageous properties such as high flexibility, ease of processability, non-toxicity, excellent charge carrier mobility and so on [1, 2]. Among the various organic materials, polymer composites of GO and rGO have been at the top of mainstream research [3]. This is due to their unmatched properties such as high flexibility, excellent electrical conductivity, solubility in most solvents

K. K. Gogoi · N. S. Das · A. Chowdhury (✉)
Organic Electronics and Sensor Laboratory, Department of Physics, National Institute of Technology Silchar, Silchar, Assam, India

R. G. Nair et al. (eds.), *Proceedings of 28th National Conference on Condensed Matter Physics*, Springer Proceedings in Physics 269,
https://doi.org/10.1007/978-981-16-5407-7_12

and so on [4]. For instance, in a recent work Sun et al. reported the RS properties of GO/polyvinyl alcohol (PVA) composites with a V_{SET}/V_{RESET} of ~ −0.75/3.0 V and $I_{ON/OFF}$ ratio ~3×10^4 [5]. Similarly, GO composites with polyimide and polystyrene polymers have also been reported by researchers for RS applications [6, 7]. Apart from GO/polymer composites, composites of rGO with polymer have also been studied for RS applications. For example, composites of rGO with polymers such as polymethyl methacrylate (PMMA) and PFCF-CHO have been found to exhibit excellent bipolar RS properties [8, 9]. In the present work, RS properties of PMMA composites with rGO that are prepared through microwave reduction of GO have been reported. Furthermore, RS properties of rGO have been tuned by varying the reduction level of GO which forms the novelty of the work.

12.2 Experimental Procedure

12.2.1 Synthesis of rGO

GO was synthesized through conventional Hummer's method in which a solution containing graphite flakes and sodium nitrate was prepared in sulfuric acid. In the next step, potassium permanganate was added to the solution which was then heated at 96 °C followed by the addition of deionized water. Finally, hydrogen peroxide was added and the solution was washed with 10% HCl and deionized water to obtain the GO product [10, 11]. The as-synthesized GO was then placed inside a microwave oven and was irradiated at three different powers (540, 720 and 900 W) for 60 s each. The products thus obtained were the desired rGO.

12.2.2 Fabrication of Devices

For the fabrication of devices, composites containing 0.1 mg each of rGO (540, 720 and 900 W) and 99.9 mg of PMMA were prepared in DMF solvent through ultrasonication. The prepared composites were spin coated onto pre-cleaned ITO at 9000 rotations per minute (RPM) for 60 s. The top aluminum (Al) electrode was then deposited through thermal evaporation technique. A schematic diagram of the fabricated device is shown in Fig. 12.1. The fabricated devices were named as $RGPC_{540}$, $RGPC_{720}$ and $RGPC_{900}$ for reduction at 540, 720 and 900 W, respectively.

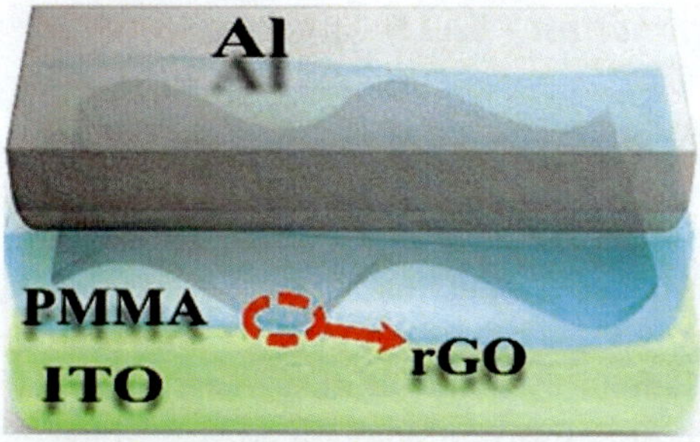

Fig. 12.1 Schematic diagram of ITO/rGO-PMMA/Al device

12.3 Results and Discussions

12.3.1 Photoluminescence (PL) Spectroscopy

Figure 12.2a shows the PL spectra of rGOs excited with a wavelength λ_{ex} of ~230 nm. From the spectra, it can be seen that all the rGOs exhibit a broad spectrum with a common peak positioned at ~396 nm which corresponds to the presence of defect states in the rGO structure that occur either during the synthesis of GO or during its reduction [12]. The defects result from the order less arrangement of the sp^3 bonded carbon atoms in the rGO structure. Also, the presence of defects in the rGO structure leads to the formation of various intermediate energy levels between its conduction and valence bands which are responsible for the broad nature of the PL spectrum [13]. Furthermore, the PL peaks of rGOs are blue-shifted as compared to that of GO which is due to the narrowing of the band gap of GO upon reduction [13]. It is also observed that with the increase in reduction level, the peak intensities decrease which is indicative of the decrease in the number of defect states.

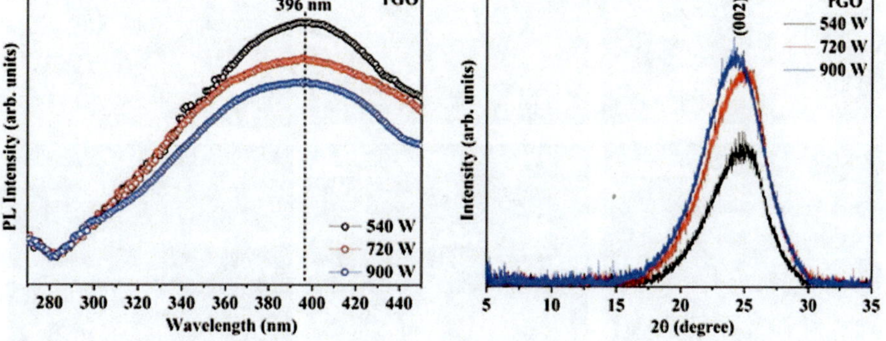

Fig. 12.2 **a** PL spectra and **b** XRD patterns of rGO reduced at 540, 720 and 900 W

12.3.2 X-ray Diffraction (XRD) Spectroscopy

Figure 12.2b shows the XRD pattern of synthesized rGOs. From the figure, it can be seen that all the rGOs exhibit a similar peak at ~25° which correspond to the reflection from (002) plane [14]. Also, no peak corresponding to GO is observed which confirms the successful reduction by microwave irradiation technique. The interplanar spacing for rGOs is calculated to be ~3.56 Å which is less than GO (9.7 Å). This reduction is due to the removal of oxygen functionalities from the basal planes of GO during the microwave irradiation process which reduces the separation between the various planes of rGO.

12.3.3 Scanning Electron Microscopy (SEM) and Transmission Electron Microscopy (TEM)

The SEM images of rGOs are shown in Fig. 12.3a–c. From the figure it can be seen that all the rGOs possess a crumpled sheet like morphology. Also, the TEM images in Fig. 12.3d–f confirm the sheet like structure of the rGOs with wrinkles throughout.

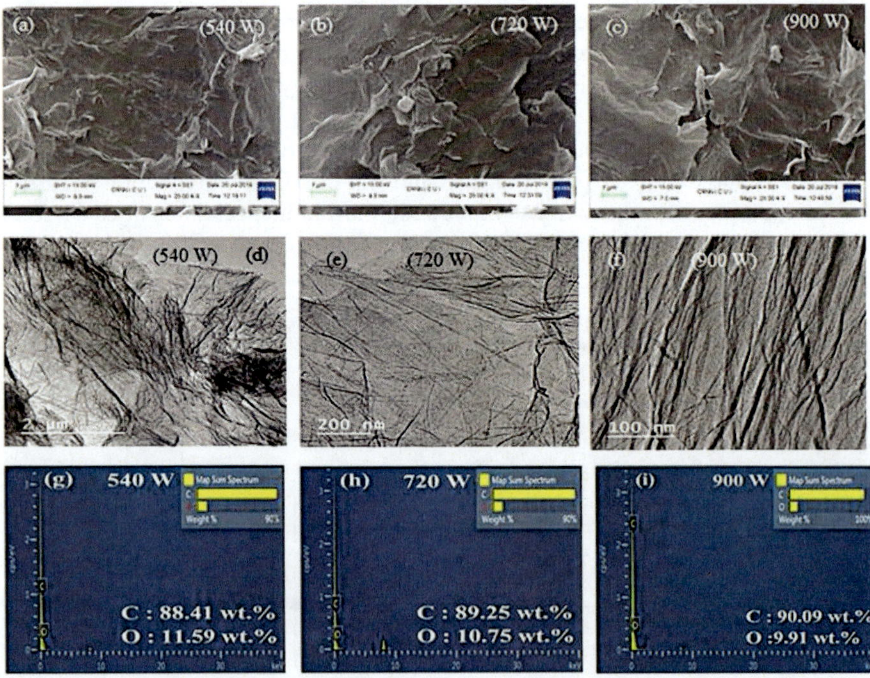

Fig. 12.3 **a–c** SEM images; **d–f** TEM images and **g–i** EDX mappings of rGO reduced at 540, 720 and 900 W, respectively

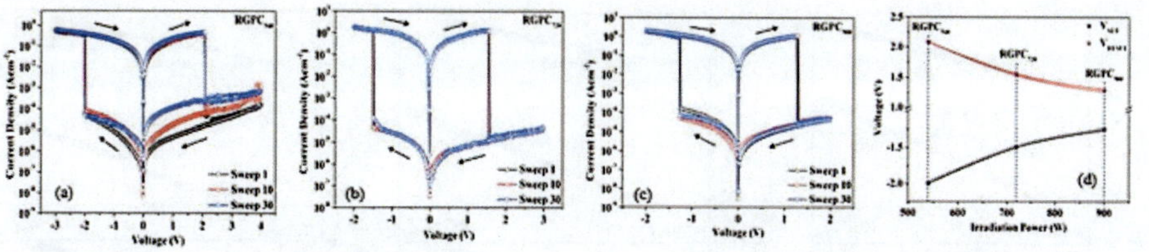

Fig. 12.4 **a–c** J-V characteristics and **d** Variation of V_{SET}/V_{RESET} with irradiation power of $RGPC_{540}$, $RGPC_{720}$ and $RGPC_{900}$ devices, respectively. Arrows in **(a–c)** represent the direction of voltage sweep

However, the wrinkles appear to get ordered upon increasing the power of microwave irradiation which may be due to the increase in ordered sp^2 bonded carbon atoms upon reduction. The energy dispersive X-ray (EDX) mappings (Fig. 12.3g–i) display a gradual decrease in the wt. % of oxygen upon increasing the power of microwave irradiation which confirms the successful tuning of the reduction levels of GO.

12.3.4 Current Density–Voltage (J-V) Analysis

The J-V characteristics of the ITO/rGO-PMMA/Al devices is shown in Fig. 12.4a–c. From the figure, it can be seen that all the devices exhibit highly stable and repeatable bipolar resistive switching properties. V_{SET}/V_{RESET} as low as $-1.25/+1.28$ V are obtained for the $RGPC_{900}$ device which increase with the decrease in the power of reduction ($RGPC_{900} > RGPC_{720} > RGPC_{540}$) as shown in Fig. 12.4d. Also, a high I_{ON}/I_{OFF} ratio of 10^4 is exhibited by all the RGPC devices.

The conduction in the RGPC devices occur due to the phenomenon of trapping and de-trapping of charge carriers in the defect sites of rGO [15]. When a negative ($-$ve) bias is applied on the Al electrode ($0 \rightarrow -V$), charge carriers from Al are initially blocked by the insulating PMMA layer which results in the accumulation of charge carriers in it. These accumulated carriers form the space charge and thus lead to a slow rise in current. Upon increasing the external $-$ve bias, the charge carriers diffuse through the PMMA layer and get trapped in the defect sites of rGO. When the defect sites get completely filled, a large number charge carriers flow from Al to ITO due to which the devices transit from the OFF to the ON state. These trapped charges remain intact even for a voltage sweep from $-V \rightarrow 0$ due to which the devices continue to remain in the ON state thereby exhibiting non-volatile characteristics. Upon altering the voltage polarity ($0 \rightarrow +V$ the charge carriers get de-trapped which destroys the trap free environment and the devices switch from ON to the OFF state. Furthermore, the decrease in V_{SET}/V_{RESET} with the increase in reduction level may be accounted to the decrease in the number of defect sites (confirmed from PL spectroscopy). This is because a smaller number of defect sites can be filled or emptied at a faster rate as compared to a larger number of defects which in turn reduces the required voltage (low V_{SET}/V_{RESET}). Finally, for a better

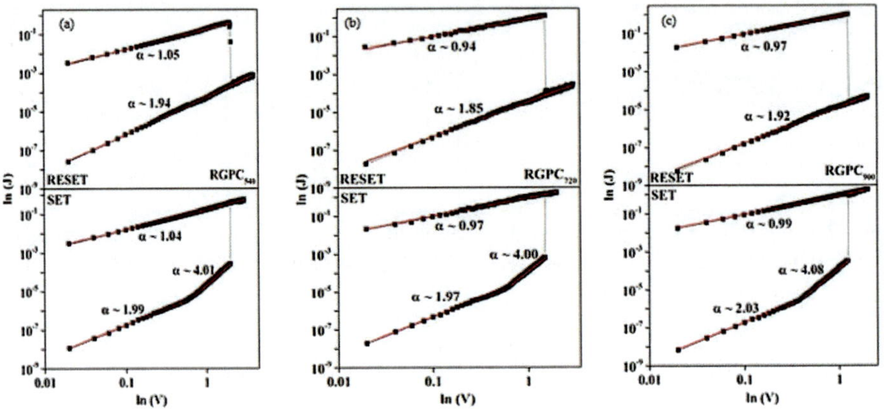

Fig. 12.5 Log–log plot of **a** $RGPC_{540}$, **b** $RGPC_{720}$ and **c** $RGPC_{900}$ devices

insight into the conduction mechanism in the RGPC devices, the J-V curves of all the devices were plotted in double log plot as shown in Fig. 12.5. From the figure, it can be seen that during the set process, all the devices initially follow the space charge limited conduction (SCLC; $\alpha \sim 2$) at low bias (OFF state) [16]. Upon increasing the external bias, the devices start obeying the trap charge limited conduction (TCLC; $\alpha \sim 4$) (OFF state) [10]. However, after switching to the ON state, the devices are found to obey the Ohmic conduction mechanism ($\alpha \sim 1$) [17]. In a similar manner, the reset process also shows the devices to obey the Ohmic conduction mechanism in the ON state and SCLC mechanism in the OFF state.

12.4 Conclusion

In summary, GO prepared through Hummer's method was successfully subjected to different levels of reduction by varying the power of microwave irradiation. The PL spectra of rGOs showed the presence of defect states in their structure which gradually decreased with the increase in the level of reduction. The XRD patterns exhibited no peak corresponding to GO which confirm a successful reduction. Furthermore, the EDX mappings displayed a reduction in the wt. % of oxygen in the rGO structures with the increase in irradiation power (rGO 540 W: 11.59 wt. %, rGO 720 W: 10.75 wt. % and rGO 900 W: 9.91 wt. %). Finally, the J-V analysis showed the presence of bipolar resistive switching properties in all the RGPC devices with a low V_{SET}/V_{RESET} of $\sim -1.25/+1.28$ V for the $RGPC_{900}$ device. Also, the devices exhibited a decrease in V_{SET}/V_{RESET} values with the increase in irradiation power. Such low-cost devices with excellent data storage properties may be used to replace the conventional Si-based memories in the future.

Acknowledgements The authors acknowledge CIF NIT Silchar, Assam and SAIF, IIT Bombay for providing the facilities of material characterization.

References

1. S. Gao, X. Yi, J. Shang, G. Liu, R.W. Li, Chem. Soc. Rev. **48**, 1531–1565 (2019)
2. L. Zhou, J. Mao, Y. Ren, S.T. Han, V.A.L. Roy, Y. Zhou, Small **14**(10), 1703126 (2018)
3. C. Tan, Z. Liu, W. Huang, H. Zhang, Chem. Soc. Rev. **44**, 2615–2628 (2015)
4. S.K. Pradhan, B. Xiao, S. Mishra, A. Killam, A.K. Pradhan, Sci. Rep. **6**, 26763 (2016)
5. Y. Sun, J. Lu, C. Ai, D. Wen, Phys. Chem. Chem. Phys. **18**, 11341–11347 (2016)
6. J.Y. Choi, H.C. Yu, J. Lee, J. Jeon, J. Im, J. Jang, S.W. Jin, K.K. Kim, S. Cho, C.M. Chung, Polymers **10**(8), 901 (2018)
7. Y. Sun, D. Wen, X. Bai, J. Lu, C. Ai, Sci. Rep. **7**, 3938 (2017)
8. B. Zhang, G. Liu, Y. Chen, L.J. Zeng, C.X. Zhu, K.G. Neoh, C. Wang, E.T. Kang, Chem. Eur. J. **17**, 13646–13652 (2011)
9. S. Vallabhapurapu, L.D.V. Sangani, M.G. Krishna, V.V. Srinivasu, C. Du, S. Du, A. Srinivasan, Mater. Today: Proc. **9**, 615–620 (2019)
10. K.K. Gogoi, A. Chowdhury, J. Appl. Phys. **126**, 025501 (2019)
11. W.S. Hummers Jr., R.E. Offeman, J. Am. Chem. Soc. **80**(6), 1339–1339 (1958)
12. C.H. Chuang, Y.F. Wang, Y.C. Shao, Y.C. Yeh, D.Y. Wang, C.W. Chen, J.W. Chiou, S.C. Ray, W.F. Pong, L. Zhang, J.F. Zhu, J.H. Guo, Sci. Rep. **4**, 4525 (2014)
13. A. Bhaumik, J. Narayan, J. Appl. Phys. **121**, 125303 (2017)
14. K.K. Gogoi, A. Chowdhury, J. Phys. Chem. C **124**, 1108–1120 (2020)
15. R. Vartak, A. Rag, S. De, S. Bhat, Appl. Nanosci. **8**, 1343–1351 (2018)
16. K.K. Gogoi, A. Chowdhury, J. Appl. Phys. **127**, 065501 (2020)
17. A.N. Aleshin, P.S. Krylov, A.S. Berestennikov, I.P. Shcherbakov, V.N. Petrov, V.V. Kondratiev, S.N. Eliseeva, Synth. Met. **217**, 7–13 (2016)

Chapter 13
Solar Light-Driven Photocatalytic Activity of CuO Nanospindles Synthesised Via Plasma-Liquid Interaction

Palash Jyoti Boruah and Heremba Bailung

Abstract This article introduces a simple single-step synthesis method of metal oxide nanoparticles using Plasma-Liquid Interaction (PLI). Due to the unique electrical, optical and catalytic properties of spindle shape copper oxide (CuO) nanoparticles, this article is focused on the synthesis of the same by generating plasma inside deionised water. SEM analysis of the nanoparticles shows the formation of spindle-shaped CuO nanoparticles arranged in flower-like structure. Photo-conductivity measurement of the CuO nanospindles indicates the generation of electron-hole pairs after illuminating light. Enhanced photocatalytic activity of the CuO nanospindles under simulated solar light is observed after adding H_2O_2 in Methylene Blue (MB) dye solution.

13.1 Introduction

In the past few decades, nanomaterials (NM) are immensely investigated due to their unique optical and electrical properties from their bulk counterparts. They serve as the bridges between the bulk material and atomic or molecular structure. Size and shape often influence the properties of metal and metal oxide nanoparticles (NP) [1]. There are many shape and size control synthesis methods such as wet chemical methods, vaporising metal in air, mechanical milling of commercial powder, magnetron sputtering in He/Ar-O_2 atmosphere, hydrothermal based methods, thermal decomposition method and solution plasma method [2]. Metal oxide NPs are significant semiconducting materials that are used in many applications (e.g. photovoltaic devices, water purifications, sensors, photocatalysis and biological applications) due to their tunable optical, morphological and charge transport properties as well as high temperature stability [3]. Primarily chemical-based or gas phase plasma is used for the synthesis of various metal oxide NPs such as TiO_2, ZnO, CuO, MnO, WO_3 etc.

P. J. Boruah (✉) · H. Bailung
Plasma Application Laboratory, Physical Sciences Division, Institute of Advanced Study in Science and Technology, Paschim Boragaon, Guwahati 781035, India

R. G. Nair et al. (eds.), *Proceedings of 28th National Conference on Condensed Matter Physics*, Springer Proceedings in Physics 269,
https://doi.org/10.1007/978-981-16-5407-7_13

In recent years, Plasma-Liquid Interaction (PLI) has gained a lot of attention in the field of nanomaterial synthesis. It offers a single-step, rapid and large scale synthesis of uniform NMs with different shape and size [4]. The formation of metal oxide nanoparticles of reactive metals such as Cu, Zn, W etc., is relatively easier than the pure metallic NPs due to the presence of strong oxidising species in the plasma zone. Saito et al. reported the control synthesis of ZnO nanoflowers and nanospheres using PLI and investigated their photocatalytic activities [5]. Cupric oxide or copper(II) oxide (CuO) is a p-type semiconductor. It has an indirect bandgap of 1.2 eV [6] and a direct bandgap of 1.85 eV [7]. By varying the structure of the material (e.g. nanosphere, nanocube, nanowire, nanospindles, nanotubes etc.), the bandgap can be modified [8, 9]. Yang et al. estimated the bandgap of flower, plate, boat and ellipsoid-like CuO nanostructures to be 1.425, 1.447, 1.429 and 1.371 eV respectively [1]. Several reports have also been published on the synthesis of CuO nanoparticles by generating plasma inside liquids. Lu et al. observed the formation of amorphous CuO NPs having size 50 nm after generating plasma in $NaNO_3$ solution [10]. When CTAB was added to the initial solution, uniform spindle-like CuO NPs having size 250–350 nm were formed. However, the synthesis of CuO nanoflowers with many sharp nanorods by generating plasma in only K_2CO_3 solution without the use of CTAB or any stabilising agents was also reported [11]. Although a few research papers have been published on the synthesis of spindle-shaped CuO NPs by PLIs, most of the properties, such as electrical, optical and catalytic etc., are yet to be investigated. The previous reports suggest the improved optical and catalytic properties of the metal oxide NPs synthesised via PLIs than the other synthesis techniques [12–14]. Therefore, plasma synthesised CuO nanospindles might also have some unrevealing properties which have to be investigated. Investigation of photocatalytic activities of CuO nanospindles will be a very interesting work to perform. This work mainly discusses the plasma generation inside deionised water to synthesise CuO nanospindles and their properties. The growth process of CuO nanospindles is also discussed.

13.2 Experimental Details and Synthesis Process

Figure 13.1a shows the schematic of the experimental setup to generate the plasma inside a liquid. A negative DC voltage of 900 V is applied to generate the plasma between two copper electrodes for 10 min. Figure 13.1b shows the generation of plasma inside deionised water. The colour of the water turns black during the experiment. After 10 min of discharge, the solution is kept in a beaker for around 24 h to settle the particles at the bottom of the beaker. The wet particles are then dried at 100 °C for 2 h to get the powdered form.

A number of electrode configurations and power supplies for the generation of plasma inside various liquids are used to synthesise CuO nanoparticles [4]. Hu et al. reported pin-to-pin electrode configuration for the synthesis of CuO NPs by generating plasma between two copper electrodes inside the NaCl solution using a pulse

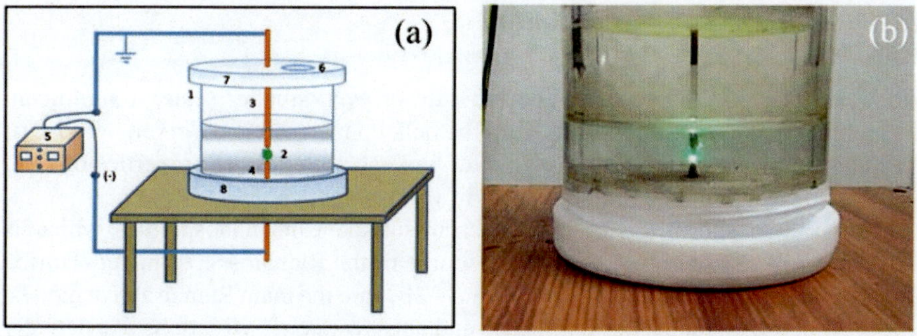

Fig. 13.1 schematic diagram of the experimental setup and (b) plasma generation inside deionised water. 1—Acrylic tube, 2—plasma, 3, 4—electrodes, 5—power supply, 6—gas outlet, 7—cover plate, 8—base plate

DC power supply [15]. Recently, a large scale synthesis of spindle-like CuO NPs using DC power supply in a pin (Pt needle) to plane (Cu foil) electrode configuration inside a solution of $NaNO_3$ mix with CTAB (Cetyl trimethylammonium bromide) was also reported [10]. The formation of Copper oxide nanorods having pointed ends by AC arc discharge inside $NaNO_3$ solution was reported by Yao et al. [16]. Saito et al. fabricated CuO nanoparticles using copper electrode by generating plasma in K_2CO_3 or citrate buffer solution [11].

To maximise the electric field at the tip of the electrodes pin-to-pin electrode configuration is used [12]. The electric field induces heat to the surrounding and vaporises the liquid medium to form a gaseous channel between the two electrodes. After reaching a specific voltage, dissociation of water vapour inside the channel take place due to the collision of energetic electrons, where the temperature increases to around 8,322 K [17]. The discharge that takes place at this high temperature is termed as thermal. It expands rapidly to the surrounding liquid and a plasma plume joining the two electrodes is formed. Since the temperature goes beyond the boiling point of the metal electrode (copper), atomic copper can easily evaporate from the electrode tips. Eventually, the active radicals generate during the discharge oxidise the copper atoms to form CuO. As the emitted particles have thermal motion, they diffused to the surrounding liquid and aggregates to form CuO NPs. The detailed mechanism of CuO nanospindles formation is discussed in our previous report [17]. Figure 13.2 summarise the mechanism of the growth process of CuO nanospindles.

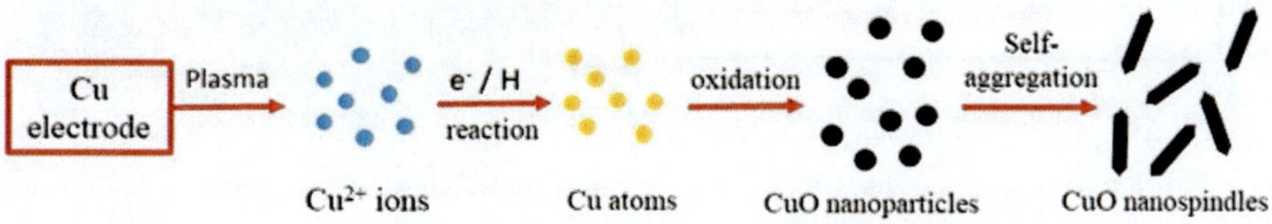

Fig. 13.2 Mechanism of the growth process of CuO nanospindles by PLI

13.3 Results and Discussion

When the size of CuO reduces to nanoscale or even smaller scale, a significant deviation of its physical properties from the bulk counterpart is observed because of the quantum-size effect. Therefore a better understanding of the properties of CuO nanoparticles is very crucial to use them in various applications.

Figure 13.2a shows the Raman spectrum of the CuO nanospindles which is obtained with a 514 nm laser excitation source of the Raman spectrometer (Horiba Jobin Vyon, LabRam HR). In CuO, the ($A_g + 2B_g$) are the main Raman active modes, where A_g and B_g denotes the symmetric and anti-symmetric vibrations respectively of the molecule with respect to the principal symmetry axis. The A_g mode is appeared at ~271 cm^{-1}, whereas B_g modes appear at ~324 and 610 cm^{-1} [16]. However, the sample shows a broadening and slight shift in the peak positions in comparison with the Raman spectrum of a single crystal CuO, which might be due to the shape and size effect of CuO NPs [18]. The XRD spectrum of CuO nanospindles shown in our previous work reported the monoclinic crystal structure of CuO [17]. Morphological property plays a vital role in the catalytic efficiency of NPs [19]. Figure 13.3b shows the SEM image of the CuO nanospindles. The nanospindles tend to form flower-like structure, as indicated by the yellow dotted circles. Nanospindles have a higher surface to volume ratio than spherical structures. Therefore, a high number of active sites to initiate photochemical reactions leads to the enhancement of the mechanism of diffusion and transportation of dye pollutants for higher photo-degradation can be expected.

The optical property of the CuO nanospindles is investigated using UV-Visible Spectroscopy, in which absorption is found to cover nearly the entire visible and UV region, as reported in our previous work [17]. Therefore CuO nanospindles could be a promising catalyst for the photo-degradation of pollutant organic dye. To determine the bandgap of the nanospindles, the following Tauc's equation is used [20].

Fig. 13.3 **a** Raman spectrum and **b** SEM image of CuO nanospindles. Yellow dotted circles indicate the flower-like structures formed by the aggregation of nanospindles

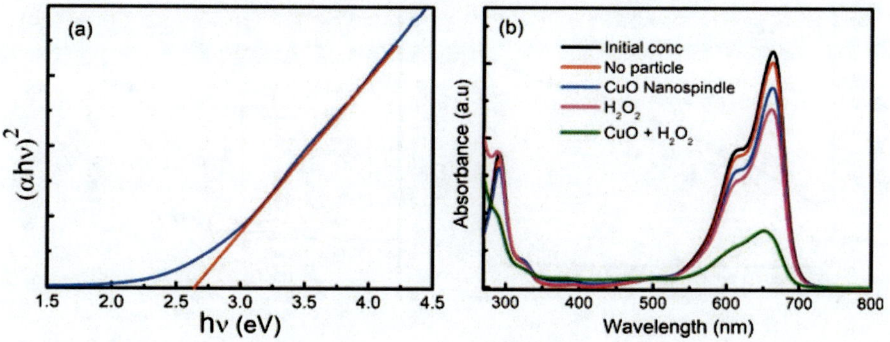

Fig. 13.4 **a** Tauc's plot for bandgap determination and **b** Photo-degradation of Methylene Blue (MB)

$$\alpha h\upsilon = A(h\upsilon - E_g)^n$$

where α, h, ν, A and E_g are the absorption coefficient, Planck's constant, frequency of the incident photon, absorption constant and optical bandgap respectively. n is an index whose value depends on the nature of transitions. For direct and indirect allowed transitions, the value of n is ½ and 2 respectively. The direct bandgap of the CuO nanospindles is calculated to be 2.64 eV, as shown in Fig. 13.4a.

To investigate the photocatalytic activity of the CuO nanospindles, photo-degradation of Methylene Blue (MB) is observed under a solar simulator (XENON DC 350). The amount of nanoparticles, initial concentration and volume of the solution are 5 mg, 20 μM and 30 ml respectively. Figure 13.4b shows the photo-degradation of MB after 6 min of solar radiations. When no particles are added the degradation is only 5% of the initial concentration. After adding CuO nanospindles and 1 M of 2 ml hydrogen peroxide (H_2O_2) separately, the degradation is increased to 15% and 24% respectively. When both H_2O_2 and the nanoparticles are added the degradation is further enhanced to 75%. The mechanism of photo-degradation can be explained as, by absorbing solar radiation, CuO creates electron-hole pairs, which can generate oxidant species such as OH^*, O_2^{*-} etc. However, the recombination probability of electron-hole pair is very high; therefore, the degradation is very low. When H_2O_2 is added, it reduces the recombination probability by scavenging the electrons and trapping the holes [21]. Hence, higher photo-degradation is observed. The following equations illustrate the mechanism of photo-degradation

$$CuO + h\upsilon \rightarrow h_{VB}^{+} + e_{CB}^{-}$$

$$H_2O_2 + h_{VB}^{+} \rightarrow OOH^* + H^+$$

$$OOH^* \leftrightarrow O_2^{*-} + H^+$$

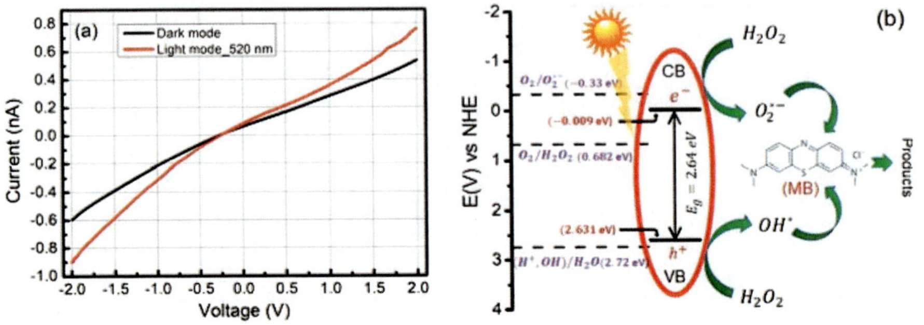

Fig. 13.5 **a** I-V curve of CuO nanospindles in the presence of light and dark, and **b** Schematic of the band alignment of CuO nanospindles along with the mechanism of photo-activation for the degradation of MB

$$H_2O_2 + e_{CB}^- \rightarrow OH^* + OH^-$$

$$OH^* + MB \rightarrow degradationproducts$$

$$O_2^{*-} + MB \rightarrow degradationproducts$$

Figure 13.5a shows the photoconductivity of the CuO nanoparticles in the dark and under light illumination (photoexcitation at 520 nm). The current-voltage (I-V) measurement is conducted in a sweep mode under ±2 V at a scan rate of 0.01 V/s. Under light illumination, the photocurrent is found to enhance, which indicates the formation of electron-hole pairs. The mechanism of photo-degradation of MB could be easily explainable by calculating the band edge alignment of CuO nanospindles relative to the Normal Hydrogen Electrode (NHE). Figure 13.5b shows the schematic of the band alignment of CuO nanospindles along with the mechanism of photo-activation for the degradation of MB. The valance band (VB) edge potential (E_{VB}) and conduction band (CB) edge potential (E_{CB}) are calculated using the expressions reported in [12, 22] and the values are −0.009 eV and 2.631 eV respectively. As E_{CB} of CuO is more positive than the standard redox potential of O_2/O_2^{*-} (−0.33 eV vs NHE), therefore the electrons at CB cannot reduce O_2 to O_2^{*-} [23]. Similarly, the holes cannot reduce H_2O to OH^*. However, the E_{CB} of CuO is more negative than the standard redox potential of O_2/H_2O_2 (0.682 eV vs NHE). Hence, the electrons at the CB can be transferred to adsorbed molecules of oxygen to produce H_2O_2. It can react with electrons to produce active OH^* radicals to degrade the dye molecules. The little degradation observed when only CuO is added to the solutions provides the evidence of MB degradation by this mechanism. When H_2O_2 is added along with CuO nanospindles, it integrate with photo-excited electrons to form $OH^*$and O_2^{*-} and effectively degrade the dye molecules.

13.4 Conclusion

The prime emphasis of the work is to provide a basic idea about the generation and interaction of plasma with liquid media to synthesise CuO nanoparticles. Plasma-liquid Interaction (PLI) offers very rapid reactions of reactive species and energetic electrons for the synthesis of nanoparticles. This method has several advantages, including low operational cost, environment friendly, no need of toxic chemical reducing agents, and, most importantly, a single-step process, consuming significantly less time to synthesise nanomaterials. Morphological analysis reveals the formation of spindle-shaped CuO nanoparticles, which have a very high tendency to form flower-like structure. The utilisation of CuO nanospindles as an efficient solar light-driven photo-catalyst is discussed. Moreover, the role of H_2O_2 on the photo-degradation of Methylene Blue (MB) along with the CuO nanoparticles is also discussed.

Acknowledgements One of the authors P. J. Boruah would like to thank DST, Govt. of India, for support under DST-INSPIRE scheme. The authors would like to thank Dr. A. R. Pal and Jyotisman Bora for helping during the photoconductivity measurement.

References

1. M. Yang, J. He, J. Colloid Interface Sci. **355**, 15 (2011)
2. X. Xu, W. Wang, W. Zhou, Z. Shao, Small Meth. **2**, 1800071 (2018)
3. L.A. Kolahalam, I.V. Kasi Viswanath, B.S. Diwakar, B. Govindh, V. Reddy, Y. L. N. Murthy, Mater. Today Proc. **18**, 2182 (2019)
4. Q. Chen, J. Li, Y. Li, J. Phys. D. Appl. Phys. **48**, 424005 (2015)
5. G. Saito, Y. Nakasugi, T. Yamashita, T. Akiyama, Appl. Surf. Sci. **290**, 419 (2014)
6. A.E. Rakhshani, Solid State Electron. **29**, 7 (1986)
7. K. Santra, C.K. Sarkar, M.K. Mukherjee, B. Ghosh, Thin Solid Films **213**, 226 (1992)
8. H. Chen, G. Zhao, Y. Liu, Mater. Lett. **93**, 60 (2013)
9. T. Jiang, Y. Wang, D. Meng, X. Wu, J. Wang, J. Chen, Appl. Surf. Sci. **311**, 602 (2014)
10. Q. Lu, X. Wang, J. Yu, F. Feng, L. Yin, Y. Kang, H. Luo, Mater. Lett. **264**, 127316 (2020)
11. G. Saito, S. Hosokai, M. Tsubota, T. Akiyama, J. Appl. Phys. **110**, 023302 (2011)
12. P.J. Boruah, R.R. Khanikar, H. Bailung, Plasma Chem. Plasma Process. **40**, 1019 (2020)
13. L. Chen, T. Mashimo, H. Okudera, C. Iwamoto, E. Omurzak, RSC Adv. **4**, 28673 (2014)
14. S. Pitchaimuthu, K. Honda, S. Suzuki, A. Naito, N. Suzuki, K.I. Katsumata, K. Nakata, N. Ishida, N. Kitamura, Y. Idemoto, T. Kondo, M. Yuasa, O. Takai, T. Ueno, N. Saito, A. Fujishima, C. Terashima, ACS Omega **3**, 898 (2017)
15. X. Hu, X. Zhang, X. Shen, H. Li, O. Takai, N. Saito, Plasma Chem. Plasma Process. **34**, 1129 (2014)
16. W.T. Yao, S.H. Yu, Y. Zhou, J. Jiang, Q.S. Wu, L. Zhang, J. Jiang, J. Phys. Chem. B **109**, 14011 (2005)
17. P.J. Boruah, R.R. Khanikar, H. Bailung, Nanotechnology **32**, 245601 (2021)
18. H. Hagemann, H. Bill, W. sadowski, E. Walker, M. François, Solid State Commun. **73**, 447 (1990)
19. Q. Zhang, K. Zhang, D. Xu, G. Yang, H. Huang, F. Nie, C. Liu, S. Yang, Prog. Mater. Sci. **60**, 208 (2014)

20. W.J. Jeyarani, T. Tenkyong, N. Bachan, D.A. Kumar, J.M. Shyla, Adv. Powder Technol. **27**, 338 (2016)
21. J. Li, F. Sun, K. Gu, T. Wu, W. Zhai, W. Li, S. Huang, Appl. Catal. A Gen. **406**, 51 (2011)
22. R.G. Pearson, Inorg. Chem. **27**, 734 (1988)
23. M. Mousavi, A. Habibi-Yangjeh, M. Abitorabi, J. Colloid Interface Sci. **480**, 218 (2016)

Chapter 14
Study of Optical Properties of Different Grades Indian Cement Samples Using Terahertz Spectroscopy

Chandan Ghorui, Koalla Rajesh, P. Naveen Kumar, and A. K. Chaudhary

Abstract The paper reports the nondestructive (NDT) evaluation of different grades cement samples using time domain terahertz (THz) techniques. The THz system was indigenously designed using LT-GaAs antenna as a source and ZnTe crystal combined with polarizers in Electro-Optic (E-O) sampling mode as a detector. We have also ascertained the optical properties such as refractive index and absorption coefficients between 0.1 and 1.0 THz range. In addition, Raman and FTIR spectroscopy of these samples were also carried out to identify the vibration bands of calcium carbonate polymorphs and important hydrated phases present in different grade cements.

14.1 Introduction

Cement is a largest industrials product that manufactured in more than 120 countries in our world. The Portland cement is one of the popular name which is further divided in two categories known as (i) Ordinary Portland Cement (OPC) (ii) Portland Pozzolana Cement (PPC).The OPC type of cement is manufactured as a powder by mixing limestone and other raw materials which consist of argillaceous, calcareous and gypsum and widely used for the construction of high-rise buildings, roads, dams, bridges, flyovers where high strength and fast setting is required. Whereas PPC contains OPC clinker and gypsum. These pozzolanic materials include volcanic ash, calcined clay or silica fumes and fly ash which makes around 15–35% of cement weight. Flyash is primarily used in PPC cement. It is used in the construction of marine structures, masonry mortars, and plastering, hydraulic structures. Besides, they are popularly used in mass concreting works, such as dykes, sewage pipes, dams, etc. PPC is also employed in all applications where OPC is also used. From the spectroscopic point of view, terahertz (THz) spectral range lies between 0.1 and 10 THz range. It bridges the gap between microwave and optical region. It is a

C. Ghorui · K. Rajesh · P. Naveen Kumar · A. K. Chaudhary (✉)
Advanced Centre of Research in High Energy Materials (ACRHEM), University of Hyderabad, Telangana 500046, India
e-mail: akcsp@uohyd.ernet.in

R. G. Nair et al. (eds.), *Proceedings of 28th National Conference on Condensed Matter Physics*, Springer Proceedings in Physics 269,
https://doi.org/10.1007/978-981-16-5407-7_14

no ionizing and noninvasive technique. Moreover, it has more than 95% reflection from the metallic surfaces and high penetration from all types of packing and organic materials such as cloth, paper, polymers etc. In addition, it is highly absorbed by water molecules. Therefore, Time domain THz spectroscopy is widely used for recording of finger print spectra of organic, polymers, packing materials, drugs and bio molecules. It is also used for cement analysis in mortar form of anhydride state and imaging to understand the development of crakes [1, 2]. Similarly, Raman spectra was used to study the calcium carbonate polymorphous and in paste form. It is also used for the study of sulfacted cement hydrates. In addition, different types of OH bonds and calcium Hydroxide, C-S-H bond formation were reported using FTIR. [3]. In this study the amounts of main chemical constituents such as Silicon oxide (SiO_2), Magnesium oxide (MgO), Sulphur Trioxide (SO_3), Aluminum oxide (Al_2O_3), Iron oxide (Fe_2O_3), Calcium Oxide (CaO) and Loss on Ignition were compared. The possible reason for variation in chemical composition and their effect on properties of cement have been discussed [4, 5].

14.2 Experimental Details

The experiment was carried out in three parts, (a) sample preparation for THz spectroscopy (b) THz generation and TDTS of cement samples (c) Raman and FTIR spectroscopy. We have selected 43 and 53 grades cement samples obtained from four different companies in our study. A small quantity of cement sample weighting ~1.7 g was subjected to a specially designed Teflon window sample holder of dimensions $6.63 \times 5.59 \times 0.75$ cm^3 for making a uniform pellet of thickness 1.82 mm. These sample are placed for spectroscopic analysis between 0.1 and 1.0 THz range. ALT- GaAs made stripe antenna was used as a THz source and the Combination of ZnTe crystal, $\lambda/4$ plate, Wollaston prism and a pair of photodiodes in Electro- optic configuration as a detector. The system was pumped by 800 nm wavelength of 140 fs pulses obtained from Ti: sapphire laser oscillator at 80 MHz repetition rate [6]. The laser beam was divided into two parts of 90:10 intensities; one part was allowed to an incident on the LT-GaAs made photoconductive antenna for the generation of THz radiation, and another part was directed to the ZnTe crystal arranged in E-O sampling mode for the detection. The translation stage was used to control the delay between the pump and probe laser beams. The path lengths of the pump and probe were equalized by doing auto-correlation. The THz waveform was mapped out by measuring the photocurrent generated by a balanced photodiode. The output of the balanced photodiode was fed to the lock-in amplifier synchronized with a function generator with a frequency of 11.573 kHz, which was used to trigger the photoconductive antenna. It helps to enhance the signal to noise ratio. The experimental data were recorded by a data acquisition program, which was developed using lab view software.

In the present experiment 1 mg solid cement samples of different grade and companies were subjected to a Raman System (Model No-HORIBA Scientific, XploRA

ONE). Where a beam of 532 nm wavelength obtained from Nd: YAG laser at power of 25mW was used as a source. It has 900 line/mm grating. The sport size of the laser beam was 1–3 μ.

14.3 Results and Discussion

The terahertz time/frequency domain spectra as shown in Fig.14.1 (a and b) clearly reveal the difference among the different grade cement samples in terms their optical path lengths. Which is also attribute to their chemical composition. The time domain data was analyzed by the refractive index and absorption coefficient of different grade materials. Figure 14.2a shows the absorption coefficient and (b) shows that the refractive index graphs of different types Indian cement. Only 0.58 ± 0.01 and 1.0 ± 0.01 absorbance band present in 53 Birla and Penna cement, these bands are absent in 43 grade Sagar and Coromondal cements.0.58 THz band is attributed to gypsum. The refractive index of different type cement materials is different in 0.1–1 THz range. Only Penna 53 grade and Coromondal 43 grade show the almost same refractive index between 0.3 and 1.0 THz range. However, the refractive indices of the Birla 53 and Sagar 43 are laying between 3.0 and 1.0 range which are the highest and lowest values of cement samples (Figs. 14.1 and 14.3).

Further, the measurement of absorption coefficients quantify the absorption properties whereas dielectric properties reveals the influence of different types of ore used in the different grade cement samples on mico. Figure 14.3 (a and b) shows the dielectric constants (real and imaginary) of all the samples. The real part It is very

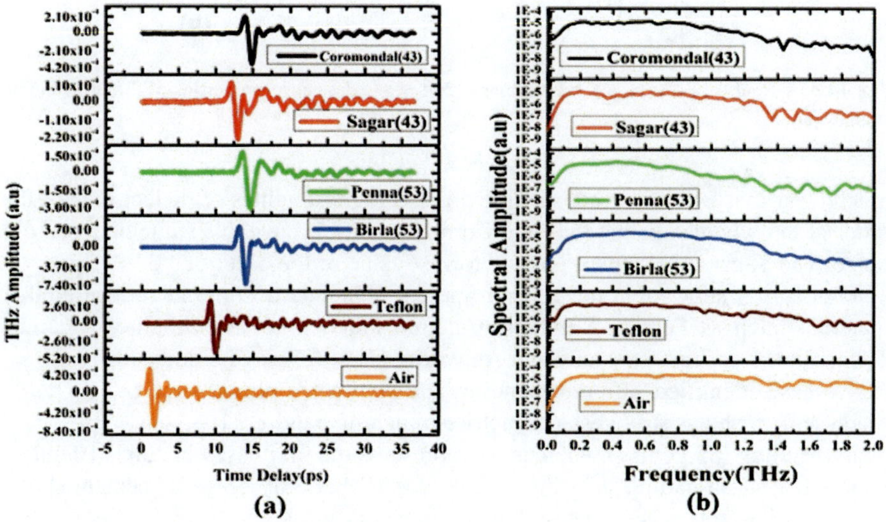

Fig. 14.1 **a** Time domain. **b** Frequency domain spectra of different grades Indian cements

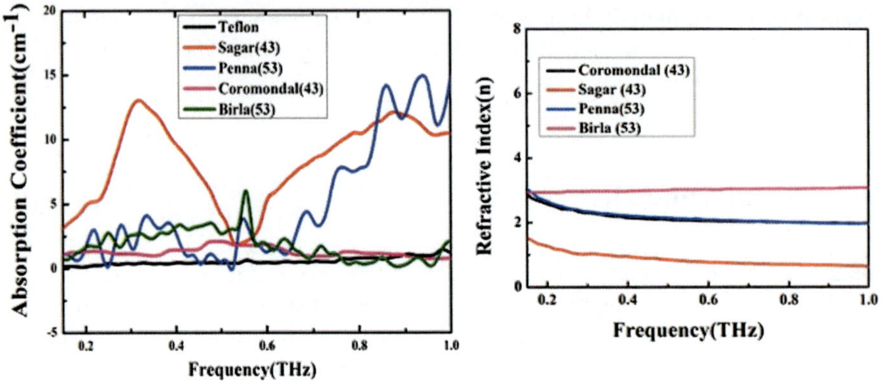

Fig. 14.2 **a** Absorption coefficients versus THz frequency. **b** Refractive Index versus THz frequency of different grade cement materials

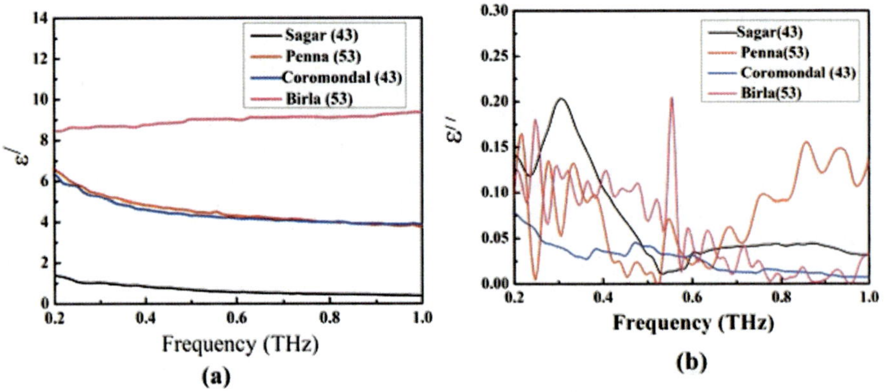

Fig. 14.3 **a** Real ($n^2 - k^2 = \varepsilon^/$). **b** Imaginary ($2nk = \varepsilon^{//}$) dielectric properties of different Indian cement samples

much clear that Birla 53 grade sample possess highest values of dielectric is of the order of 8.5 whereas Penna (53) and Coromondal (43) have this value between 6-5 range. The Sagar (43) cement possess lowest value of the order of 1.3.

Figure 14.4 shows that the Raman Spectra of standard Portland cement clinker materials such as Ferrous Sulphate, Calcium langbeinite, Aphthitalite, Arcanite, Thenardite, Gypsum, Bassanite and Anhydrite [7–9] (Figs. 14.5 and 14.6).

We have identified different bending and symmetry phases of $\boldsymbol{CO_3^{2-}}$. The 43 grade cement shows strongest symmetric strenching band ν_1 (1085 cm^{-1}), ν_2 out of phase bending (853 cm^{-1}) whereas ν_4 represents the in phase bending(710 cm^{-1}) which is weak in nature. It is interesting to note that only 53 garde cement shows two bends i.e. ν_2 that is strongest one and ν_1 a very weak.

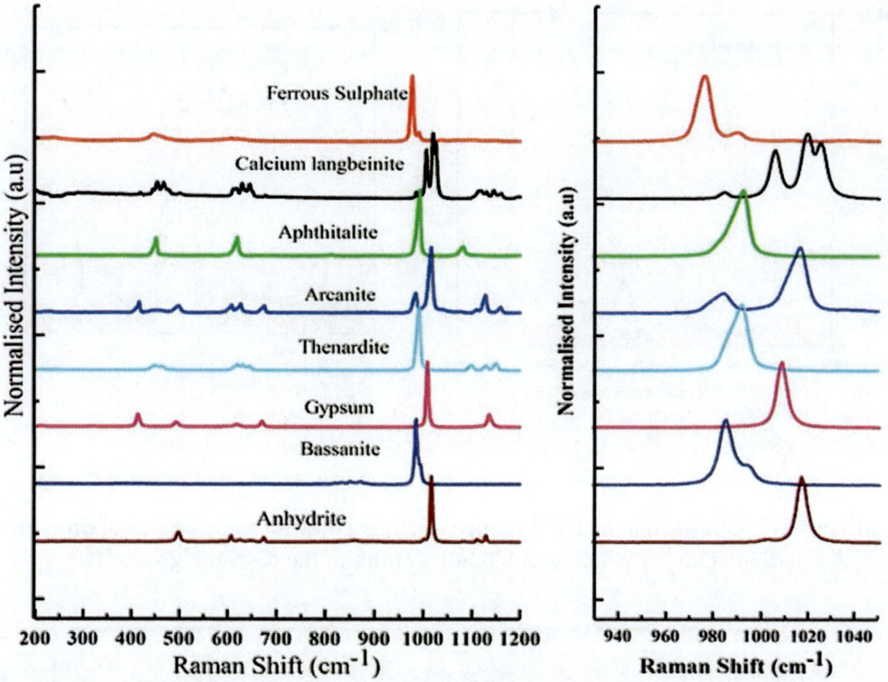

Fig. 14.4 Typical Raman Spectra of Portland cement clinker

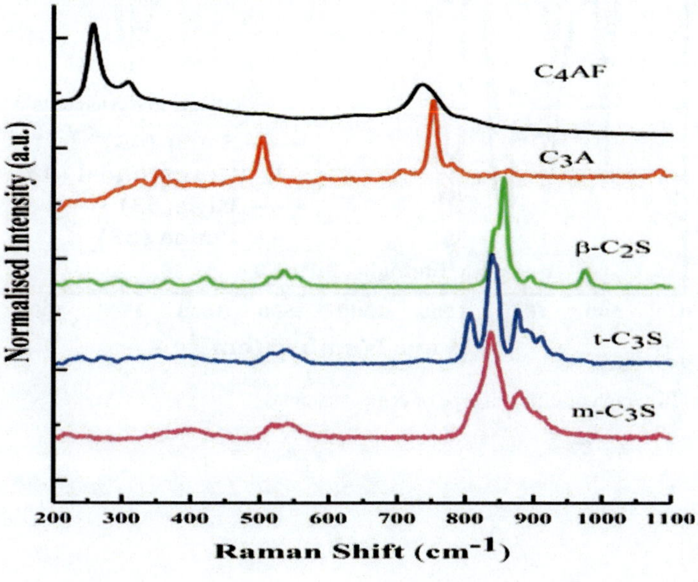

Fig. 14.5 Raman spectra from standard sample of monoclinic-C_3S, triclinic-C_3S, β-C_2S, C_3A and C_4AF

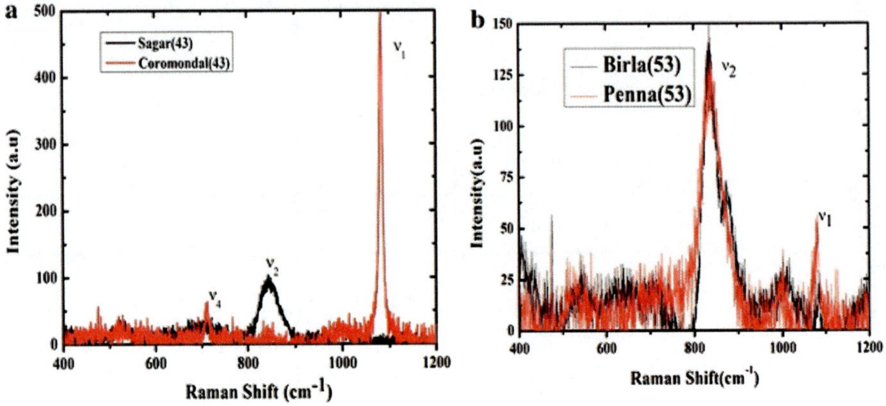

Fig. 14.6 **a** Raman Spectra of 43 grade. **b** 53 grade different indian cement samples

Figure 14.7 shows that the FT-IR bands of clinker and hydrated phases of different types of Indian cements which was already getting in the research group [10].

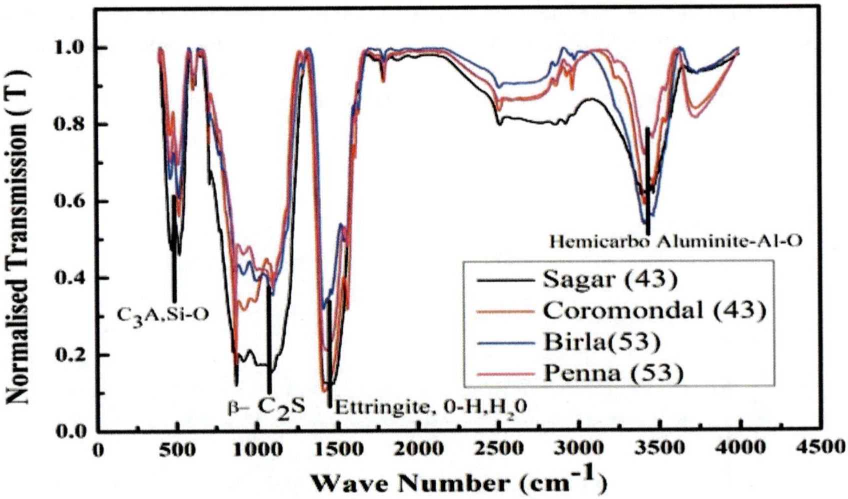

Fig. 14.7 FTIR Graph of different type of cement samples

14.4 Conclusions

We have successfully recorded the time domain THz , Raman and FTIR spectra of 43 and 53 grades Indian cements in solid powder form. In addition, absorption coefficients and Refractive index and dielectric properties in THz region were also ascertained. Finally, CO_3^{2-} stretching symmetric, in and out of phase bending is also identified.

Acknowledgements The authors would like to express their thanks to the funding agency Defense Research and Development Organization (DRDO), Ministry of Defence, Govt. of India for financial support under the grant No. DRDO/18/1801/2016/01038: ACRHEM-Phase-III and the Director of ACRHEM for all types of administrative support. Finally, the authors would like to express special thanks to Prof. Soma Venugopal Rao, ACRHEM for supplying the Raman spectra of these cement samples that is used in our experiment.

References

1. M. Tonouchi, Cutting-edge terahertz technology. Nat. Photon. **1**, 97–105 (2007)
2. J. Dash, S. Ray, K. Nallappan, S. Sasmal, B. Pesala, Non-destructive inspection of internal defects in concrete using continuous wave 2D terahertz imaging system, in *2013 38th International Conference on Infrared, Millimeter, and Terahertz Waves (IRMMW-THz)*. IEEE (2013). https://doi.org/10.1109/IRMMW-THz.2013.6665690
3. A. Abina, U. Puc, A. Jeglic, Aleksan, Applications of terahertz spectroscopy in the field of construction and building materials. Appl. Spectrosc. Rev. **50**,279–303 (2015)
4. H.-Y. Kim, S.J. Oh, C. Joo, D. Kang, Study on optical properties of nano-cement mortar using THz/sub-mm waves, in *19th World Conference on Non-Destructive Testing 2016*
5. J. Harasymiuk, A. Rudzinski, Old dumped fly ash as a sand replacement cement composition. Buildings **10**, 67 (2020)
6. D. Ganesh, A.K. Chaudhary, M. Venkatesh, Temperature-dependent terahertz spectroscopy and refractive index measurements of aqua-soluble and plastic explosives. Appl. Opt. **57**, 8743 (2018)
7. G. Renaudin, R. Segni, D. Mental, J.M. Nedelec, F. Leroux, C. Tavoit-Gucho, A Raman study of the sulfated cement hydrates: ettringite and monosulfoaluminate. J. Adv. Concr. Technol. **5**(3), 299–312 (2007)
8. Y. Yue, J. JingWang, P.A. Muhammed Basheer, J.J. Boland, Y. Bar Characterisation of carbonated portland cement pasted with optical fibre excitation Raman spectroscopy. Constr. Build. Mater. **135**, 369–376 (2017)
9. F. Liu, S.M. Asce, Z. Sun, C. Qi, Raman spectroscopy study on the hydration behaviors of portland cement pastes during setting. J. Mater. Civ. Eng. **27**(8) (2014)
10. M. Horgnies, J.J. Chen, C. Bouillon, Overview of the use of Fourier transformed infrared spectroscopy to study cementitious material, in *Conference: 6th International Conference on Computational Methods and Experiments in Materials Characterization, WIT Transactions on Engineering Sciences*, vol. 77 (2013). https://doi.org/10.2495/MC130221

Chapter 15
Synthesis and Estimation of Some Optical Properties of Fe_2O_3 Doped Bi_2O_3 Thin Film Fabricated by Eccentric Sol-Gel Route

A. K. Sahoo, D. Dwibedy, and Manas R. Panigrahi

Abstract Bismuth iron oxide is a potential material for solar cell application due to its small bandgap (~3.3 eV) and shows high-efficiency photocatalytic activity. Thus, it was attempted to synthesize the material by an unusual route. This method is a cost-effective method in comparison to the available methods like solid-state reaction, traditional sol-gel method, wet chemical method etc. It was observed that the material synthesized in a singular phase without signature of any impurity phases. The synthesized materials were annealed at 50 and 100 °C. The different optical parameters of the annealed films were estimated. Parameters like band gap, energy loss function, transition strength, optical conductivity, optical constants etc. were estimated for the film and observed to be a suitable candidate for industrial applications in photovoltaics.

15.1 Introduction

Bismuth Oxide emerges to be an important potential material for different types of application as a semiconductor material, photocatalysis due to its excellent properties including high bandgap, high refractive index, remarkable photoluminescence [1]. Bismuth oxide is also involved in various fields like solid oxide fuel cell, different types of gas sensors, superconductor materials of high-temperature, catalysis and ceramic materials. The high bandgap results to be produced pair of electron and hole when subjected to a photon beam of equal and greater energy, free radicals are produced that undergo secondary reactions [2]. Many polymorphic forms are identified for the bismuth oxide and the stable phases are α-Bi_2O_3 (monoclinic), exist at room temperature and δ-Bi_2O_3(FCC) exist above 730 °C. β-Bi_2O_3(tetragonal) and Υ- Bi_2O_3(BCC) are the intermediate metastable phases that appear during cooling of δ phase around 650 and 640 °C respectively. Moreover, some other polymorphic

A. K. Sahoo · D. Dwibedy · M. R. Panigrahi (✉)
Department of Physics, Veer Surendra Sai University of Technology (VSSUT), Burla 768018, India
e-mail: mrpanigrahi@vssut.ac.in

R. G. Nair et al. (eds.), *Proceedings of 28th National Conference on Condensed Matter Physics*, Springer Proceedings in Physics 269,
https://doi.org/10.1007/978-981-16-5407-7_15

phases like ω (triclinic) and ε (orthorhombic) can be obtained under special conditions [3]. Bismuth is known to be non-toxic, and non-carcinogenic, and also, it is valuable in electronics application, ceramic production and a good element having application in superconductors due to the strong polarizability of Bi^{3+} cations [4]. Due to its various applications in the optics related fields and many different fields, the search for optical characteristics of a material is growing day by day. It is very important to correctly establish optical constants as the quantitative optical behaviour of the material, primarily optical absorption, is very important and helpful for a good understanding of the electronic structure of the material [5]. A very important aspect is the complex dielectric function, since many of the optical properties like interband transition strength, SELF and VELF can be extracted from this function.

In this work, the optical properties of thin films of BFO annealed at 50 and 100 °C (synthesized by eccentric sol-gel method) were studied in the wavelength range (300–850) nm. This study also included the study of optical dielectric behaviour, interband transition strength, SELF and VELF. Sol-gel is emerged to be a successful technique for the generation of samples with a strong homogeneity but due to some advantages over conventional sol-gel techniques, an eccentric sol-gel was chosen, such as low cost, energy-saving, simpler manufacturing route and much smaller space is sufficient for the completion of the sample preparation process [6]. This method is therefore adopted as a prospective, affordable and reliable route for material synthesis.

15.2 Materials and Methods

An eccentric sol-gel method was used to prepare the Bismuth ferric oxide (BFO) thin film. Here, at first, Bi_2O_3 (Loba Chem, 99% pure), Fe_2O_3(Loba Chem, 99% pure), acetic acid were taken in a 250 ml beaker and a lump-free paste was obtained by stirring continuously for 30 min using a glass rod. Then hydrolysis was done to the obtained lump-free paste by dropwise addition with 100 ml of DI water, taken in another 250 ml beaker, followed by a continuous stirring with the glass rod. Later on, the stirring process was continued with a magnetic stirrer machine (IKA C-MAG HS 4) at temperature 120 °C and 230–250 rpm until a clear solution was obtained. Then, a crucial step was taken for the solution to be refluxed which followed the addition of 1 M of nitric acid with 100 ml of DI water and added to the solution dropwise, also the temperature was raised to 150 °C and the rpm to 300 but later, the rpm was decreased up to 150 to avoid sample loss from the beaker. At last, after 5 h, the BFO gel was obtained and after being cool for one day, the obtained gel was deposited on glass slides by doctor's blade technique. The thin films of BFO were fabricated and annealed at two different temperatures 50 and 100 °C. The fabricated thin films were subjected to UV–Vis spectrophotometer (Shimazdu-2540) for optical characterization.

15.3 Results and Discussions

The absorption spectra which was recorded as a function of wavelength with UV vis spectrometer in a scan range 300–850 nm, for both the thin films of BFO annealed at 50 and 100 °C was shown in Fig. 1a, which shows that the curve of both the thin films annealed at 50 and 100 °C follow almost a similar pattern i.e. the absorbance curves decreases sharply in the UV range and then slowly increases in the visible range up to ~760 nm after that again decreases. This behaviour may be due to the ratio (75:25) of the amount of reacting materials, Bi_2O_3 and Fe_2O_3 respectively. The thin film of BFO annealed at 50 °C displays greater absorption throughout the visible region and also the highest value attained. The various fundamental optical parameters can be estimated from absorption spectra.

The absorption coefficient (α) was estimated as given in (15.1)

$$\alpha = \frac{1}{d} ln \frac{1}{A} \tag{15.1}$$

where A is absorbance of the film, d is film's thickness [7]. The photon energy for each wavelength is calculated from Planck's relation and is given in (15.2)

$$\boldsymbol{E} = \boldsymbol{h}\boldsymbol{v} = \frac{hC}{\lambda} \tag{15.2}$$

The transmittance percentage (T) is related to absorbance by the relation given below,

$$\boldsymbol{T} = \mathbf{10}^{(-A)} * 100 \tag{15.3}$$

Similarly, the reflectance (R) is calculated by the equation,

$$\boldsymbol{T} = \left(1 - R^2\right) exp(-\alpha d) \tag{15.4}$$

where R is reflectance and α is coefficient of absorption. Fig. 15.1b shows the plot of transmittance versus for both the thin films of BFO annealed at 50 and 100 °C. The transmittance curve for both the thin films, after a sharp increase up to ~400 nm, decreases slowly in the visible region up to ~750 nm. Fig. 15.1c shows the reflectance versus λ curve for both the thin films which shows that the curves decrease in the visible range except ~760–850 nm region where a small increase of the curves is observed. The thin film of BFO annealed at 100 °C has greater reflectance in visible region. Fig. 15.2a, b show the variation of the refractive index and coefficient of extinction as a function of wavelength for both the BFO thin films respectively. The film's extinction coefficient(κ) is derived from Swanepoel method as

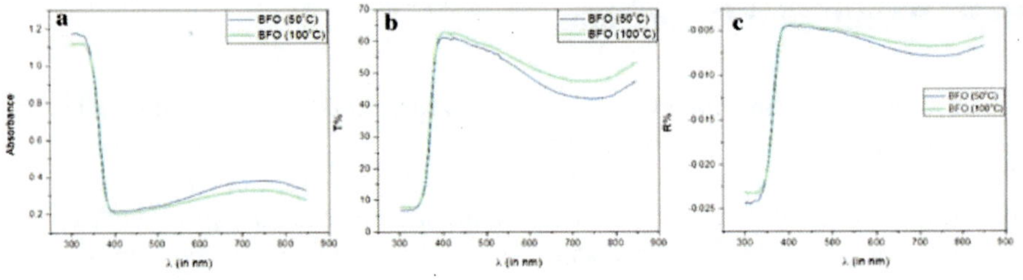

Fig. 15.1 Absorbance (**a**), transmittance (**b**) and reflectance (**c**) versus λ for both the BFO thin films annealed at 50 and 100 °C

$$\kappa = \frac{\alpha\lambda}{4\pi} \tag{15.5}$$

where α is coefficient of absorption and λ is wavelength [8]. Also, the refractive index and the transmittance percentage (T) are related [9] by (15.6)

$$\eta = \frac{1}{T} + \sqrt{\frac{1}{(T-1)}} \tag{15.6}$$

In Fig. 15.2a, both the curves rapidly decrease in ~330–390 nm of the wavelength region, which means to be normal dispersion and increasing very slowly up to ~740 nm for anomalous dispersion hence light absorption can be achieved for a long region of the wavelength in visible range i.e. from ~390–740 nm. The Extinction coefficient is an intrinsic property that is dependent upon the chemical composition and structure and measurement of light loss or light absorb at a particular wavelength due to scattering and absorption. The Fig. 15.2b shows an increase in extinction coefficient w.r.t. **λ** in the visible region signifies that these films become opaque in this region and the 50 °C annealed thin film is, therefore, more opaque [10]. The optical conductivity (σ_{opt}) is given by

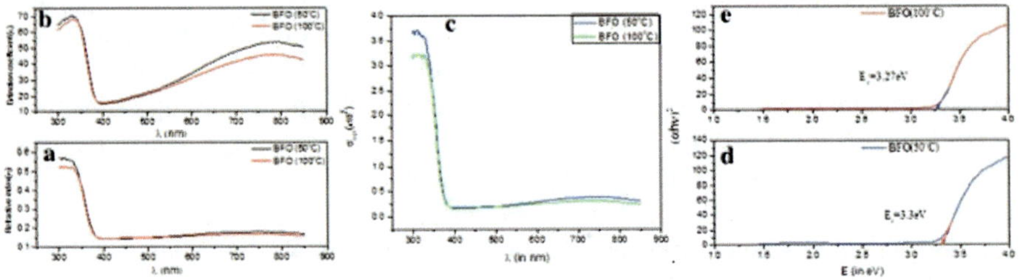

Fig. 15.2 Refractive index (**a**), extinction coefficient (**b**), optical conductivity (**c**) versus λ and bandgap estimation(D&E) for both the thin films of BFO annealed at 50 and 100 °C

$$\sigma_{opt} = \frac{\alpha \eta c}{4\pi} \tag{15.7}$$

where α is absorption coefficient, η is refraction index. Fig. 15.2c shows the curve of optical conductivity versus λ, where a sudden decrease of the curve can be seen up to ~395 nm and then the curve for both the thin films increases very slightly in the entire visible range. Figure 15.2d, e show the curve of $(\alpha h\nu)^2$ versus E and the bandgap is estimated to be 3.3 eV and 3.27 eV for the 50 and 100 °C annealed thin films respectively, by Tauc equation [7]

$$\alpha h\nu = B\left(h\nu - E_g\right)^n \tag{15.8}$$

where B is a constant, that describes the process of optical absorption, i.e. n = 1/2 for direct allowed transitions. The complex dielectric constant (ε) which is an inherent feature, can be defined [11] as

$$\varepsilon = \varepsilon_1 + i\varepsilon_2 \tag{15.9}$$

where

$$\varepsilon_1 = \eta^2 - \kappa^2 \tag{15.10}$$

and

$$\varepsilon_2 = 2\eta\kappa \tag{15.11}$$

The real dielectric constants (ε_1) is connected with the refraction property (reduction of the velocity of light inside the material). The imaginary dielectric constants (ε_2) measures the amount of energy absorbed due to the dipole electrical field. Any solid material's polarizability is directly related to its dielectric constant. It gives us details on the electronic quantity of strongly effective optoelectronic products for designing. Fig. 15.3a shows that with an increase in energy for both BFO films in the energy spectrum ~1.57–3.25 eV, ε_1 increases. This means an increase in extinction and electronic conversion to conduction band from valance band in the material. The electronic transition makes electronic collisions that can be elastic or inelastic, resulting in an increment in dielectric constant. In Fig. 15.3b, the ε_2 curve decreases at first but increases rapidly from ~3.25 eV energy for both the thin films of BFO. The value of dielectric of 50 °C annealed thin films is greater in both low and high energy level. The atoms inside the material are affected by the temperature factor that results in a change in their position and orientation, and hence the medium inside the material is also changed. Therefore, the optical parameters vary with temperature variation. Now, using ε_1 and ε_2 via the relation (15.12), the interband transition strength (Jcv) can be derived [5].

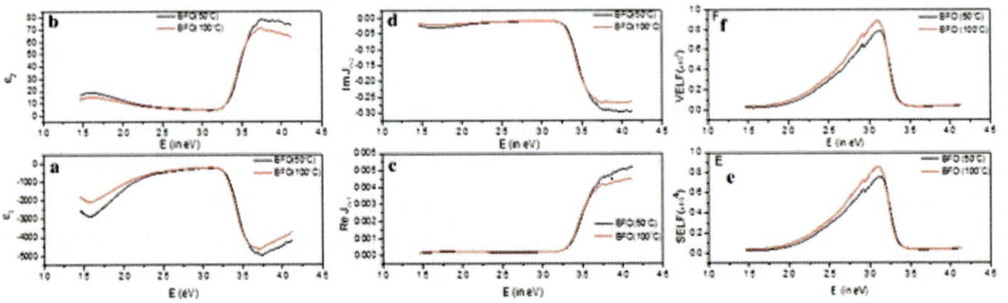

Fig. 15.3 Real and imaginary part of Dielectric constants (**a** and **b**) and Interband transition strength (**c** and **d**), SELF (**e**) and VELF (**f**) versus E for thin film of BFO annealed at 50 and 100 °C

$$J_{cv} = J_{cv1} + J_{cv2} = \frac{m_0^2 4\pi^2}{e^2 h^2}\frac{E^2}{2}(\varepsilon_2 + i\varepsilon_1) \tag{15.12}$$

Here, m_o is electron mass. The interband transition strength is important to be accounted for the possibility of an electron transition among the occupied valance band and the unoccupied conductive band. The considered optical transition is of two types, first bipolar, involving both the carriers, holes and electrons, occurs among the conduction band and the valance band, secondly, unipolar, includes just one kind of carrier and takes place either within the conduction or valance band. The excitation of electrons occurs due to the interaction of the photon with the material. These excited electrons jump into unoccupied energy level of conduction band including the collective excitation of valance electrons [12]. The curve of the ReJ_{cv1} (Fig. 15.3c) increases very quickly from the energy ~3.17 eV for both the thin films but the curve of the ImJ_{cv2} (Fig. 15.3d) decreases rapidly also from ~ 3.17 eV. This means to be the most of the high absorption occurs and results in the enhancement of the excitation of the electrons and jumping from the valance band to conduction band. Surface energy loss function (SELF) and volume energy loss function (VELF) [11] is also calculated using the ε_1 and ε_2, as given below in (15.13) and (15.14),

$$\text{surface} - \text{Im}\left(\frac{1}{1+\varepsilon}\right) = \frac{\varepsilon_2}{\left((\varepsilon_1+1)^2+\varepsilon_2^2\right)} \tag{15.13}$$

$$\text{Volume} - \text{Im}\left(\frac{1}{\varepsilon}\right) = \frac{\varepsilon_2}{(\varepsilon_1+\varepsilon_2)^2} \tag{15.14}$$

The highest value of Surface and Volume energy loss function refer to the energy of absorption. Thanks to the interband transition which takes place at 3.14 eV for SELF and 3.15 eV for VELF of the 50 °C annealed thin film, also 3.10 eV for SELF and 3.09 eV for VELF of the thin film annealed at 100 °C. The peaks at these energies correspond to be the highest extinction that takes place at this energy and the energy losses owe to the crossover of the first valance electron to the conduction band in the material of the films. The optical parameters discussed above gives an idea about the application of the synthesized material for energy harvesting for solar cell (Table 15.1).

Table 15.1 Estimated parameters of BFO thin films annealed at 50 °C and 100 °C

Annealing temperature (°C)	η	κ	E_g (eV)	ε_1	ε_2	ReJ_{cv1}	ImJ_{cv2}	σ_{opt}	SELF	VELF
50	0.57	54.01	3.3	−754.66	78.48	5.19×10^{-3}	−0.022	3.7×10^{7}	7.63×10^{-5}	7.88×10^{-5}
100	0.52	46.1	3.27	−644.31	70.87	4.49×10^{-3}	-0.015	3.2×10^{7}	8.64×10^{-5}	10^{-5}

15.4 Conclusion

The thin films of BFO annealed at 50 and 100 °C were synthesized by the eccentric sol–gel method. The important optical parameters like refractive index(η) and extinction coefficient (κ) etc. were calculated using the data of the absorption spectra from UV vis. Also, the calculation of the complex dielectric function and by using this, estimation of interband transition strength (J_{cv}), surface and volume energy loss function was included in this work.

References

1. A. Hernández-Gordillo, J.C. Medina, M. Bizarro, R. Zanella, B.M. Monroy, S.E. Rodil, Ceram. Int. **42**, 11866 (2016)
2. H. Oudghiri-Hassani, S. Rakass, F. T. Al Wadaani, K. J. Al-ghamdi, A. Omer, M. Messali, M. Abboudi, J. Taibah Univ. Sci. **9**, 508 (2015)
3. C.L. Gomez, O. Depablos-Rivera, P. Silva-Bermudez, S. Muhl, A. Zeinert, M. Lejeune, S. Charvet, P. Barroy, E. Camps, S.E. Rodil, Thin Solid Films **578**, 103 (2015)
4. A. Azuraida, M.K. Halimah, A.A. Sidek, C.A.C. Azurahanim, S.M. Iskandar, M. Ishak, A. Nurazlin, Chalcogenide Lett. **12**, 497 (2015)
5. M.A. Nattiq, J. Basrah Res. **40**, 28 (2014)
6. M. Devi, M.R. Panigrahi, IOP Conf. Ser. Mater. Sci. Eng. **653**, (2019)
7. M. Devi, M.R. Panigrahi, U.P. Singh, J. Mater. Sci. Mater. Electron. **26**, 1186 (2014)
8. M.R. Panigrahi, M. Devi, AIP Conf. Proc. **1832**, (2017)
9. F.H.A. Bouabellou, IOSR J. Eng. **3**, 21 (2013)
10. R. Chauhan, A.K. Srivastava, M. Mishra, K.K. Srivastava, Integr. Ferroelectr. **119**, 22 (2010)
11. A.I. Ali, J.Y. Son, A.H. Ammar, A. Abdel Moez, Y.S. Kim, Results Phys. **3**, 167 (2013)
12. M.F. Al-mudhaffer, J. Basrah Res. **36**, (2010)

Chapter 16
Study of Nonlinear Hybrid Optomechanical System Containing Quantum Dot: Possible Applications

Vijay Bhatt, Sabur A. Barbhuiya, Pradip K. Jha, and Aranya B. Bhattacherjee

Abstract We theoretically study a hybrid system containing of a single quantum dot (QD) embedded in a solid state microcavity which interact with the deformation potential associated with the lattice vibration and quantized cavity mode. We investigate optical bistability, mechanically induced absorption (MIA) and Fano resonance in the present system. We find that the bistability can be controlled by the QD-cavity mode coupling. We further show that the normalized power transmission displays anomalous dispersion indicating that the system can be used to generate slow light. This study also demonstrate the possibility of using the system as an all optomechanical Kerr switch.

16.1 Introduction

Quantum optomechanics has increasingly emerged as a major field of study in the recent past. In cavity optomechanics, which typically consists of a mechanical oscillator and an optical cavity, the interaction of photons and phonons through radiation pressure is exploited [1–4]. Solid state based optomechanical device has been studied earlier [5].

A single quantum dot in a solid state micro-cavity interacting with the quantized mode of light and the deformation potential associated with the vibrations of the lattice is studied in the present article. The occurrence of mechanically induced absorption (MIA), optical bistability, and Fano resonance and their dependency on exciton-pump detuning, exciton-lattice vibration coupling, and mode coupling of exciton-cavity is studied in detail.

V. Bhatt · P. K. Jha (✉)
Department of Physics, DDU College, University of Delhi, New Delhi 110078, India

S. A. Barbhuiya · A. B. Bhattacherjee
Department of Physics, Birla Institute of Technology and Science, Hyderabad Campus, Pilani, Hyderabad 500078, India

R. G. Nair et al. (eds.), *Proceedings of 28th National Conference on Condensed Matter Physics*, Springer Proceedings in Physics 269,
https://doi.org/10.1007/978-981-16-5407-7_16

We notice that the appearance of bistability can be effectively tuned by the QD-cavity mode coupling. The use of this hybrid framework in quantum information processing platforms is strongly suggested by these findings.

16.2 Theory and Model

In Fig. 16.1, the model proposed in this article is shown. We consider a single quantum dot (QD) inside a semiconductor micro cavity. The micro-cavity can be fabricated by a set of distributed Bragg reflectors (DBR). Light confinement along the longitudinal and transverse direction in the DBR can be achieved by known techniques [6].

The complete Hamiltonian is written in the rotating frame of the pump frequency ω_p, as,

$$\begin{aligned}\widehat{H}_{\mathrm{rot}} = & \hbar\Delta_p\sigma_z + \hbar\Delta_c\mathbf{a}^+\mathbf{a} + \hbar\mathbf{g}_0(\sigma_+\mathbf{a} + \sigma_-\mathbf{a}^+) - i\hbar\mathbf{E}_p(\mathbf{a}^+ - \mathbf{a}) \\ & - i\hbar\mathbf{E}_s(\mathbf{a}e^{i\delta t} - \mathbf{a}^t e^{-i\delta t}) + \sum_{\vec{k}} \omega_{\vec{k}}\left(\mathbf{b}_{\vec{k}}^+\mathbf{b}_{\vec{k}}\right) + \hbar\sigma_z \sum_{\vec{k}} \mathbf{M}_{\vec{k}}\left(\mathbf{b}_{\vec{k}} + \mathbf{b}_{\vec{k}}^+\right),\end{aligned} \tag{16.1}$$

where, first term is the energy of the exciton and $\Delta_p = \omega_{exc} - \omega_p$ denotes pump-exciton detuning. Second term is the energy of cavity photon and $\Delta_c = \omega_c - \omega_p$ denotes cavity-pump detuning. The third term is the interaction of quantum dot and cavity. Here g_0 is the coupling interaction of QD and cavity. Fourth term is the interaction energy of external pump with cavity where, E_p is the external pump strength. The fifth term is the interaction between signal laser and cavity. Here E_s is the strength of the signal laser and$\delta = \omega_p - \omega_s$ is the signal-pump detuning. The sixth term is the energy of vibrational phonon with frequency $\omega_{\vec{k}}$. The last term is the coupling between phonon and exciton and the coupling strength is denoted by $M_{\vec{k}}$.

The corresponding Heisenberg-Langevin equation is now obtained as

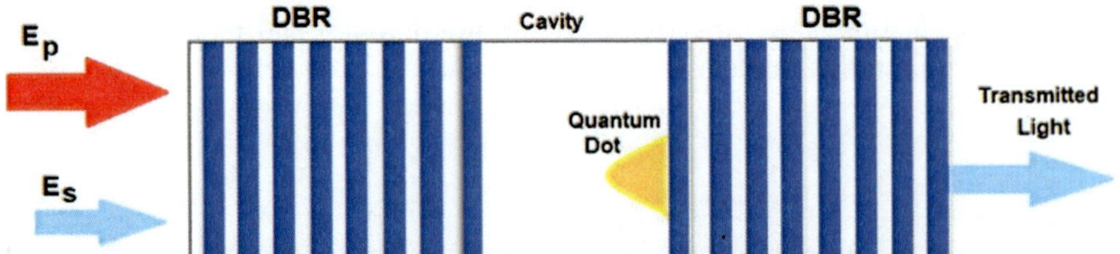

Fig. 16.1 The cavity of the DBR mirrors is shown in the figure. With the cavity, the quantum dot communicates. The GaAs and AlGaAs layers are depicted by white and blue coloured strips

$$\dot{\sigma}_z = -\Gamma_1(\sigma_z + 1) - ig_0(a\sigma_+ - a^+\sigma_-) \quad (16.2)$$

$$\dot{\sigma}_- = -(\Gamma_1 + i\Delta_p)\sigma_- - iq\sigma_- + 2ig_0 a\sigma_z + F_n, \quad (16.3)$$

$$\dot{a} = -(i\Delta_c + k_c)a - ig_0\sigma_- + E_p + E_s \boldsymbol{ae}^{-i\delta t}, \quad (16.4)$$

$$\ddot{q} + \Upsilon_q\dot{q} + 2\eta\omega_q^3\sigma_z - \omega_q^2 q = 0, \quad (16.5)$$

where, $\eta = \sum_{\vec{k}} \frac{M_k^2}{\omega_k^2}$ is the factor of Huang-Rhys which denotes the coupling of exciton-phonon.

The Spontaneous emission rate of exciton isΓ_1 and dephasing rate is given by Γ_2. Also q is the position operator of phonon. F_n is the noise operator with zero mean. Photon decay rate is given by k_c and Υ_q is the phonon decay rate. We are looking at the mean response to the signal field of the coupled device here, and thus we do not consider quantum fluctuations. Also we neglect the entanglement between the degree of freedom of phonon and exciton.

The linear optical susceptibility is deduced as,

$$\chi_{eff}^{(1)} = \frac{\sigma_+}{E_s}. \quad (16.6)$$

We deduce nonlinear optical susceptibility in a similar way as:

$$\chi_{eff}^{(3)} = \frac{\sigma_-}{3E_s^* E_p^2} \quad (16.7)$$

The transmitted output field through the coupled device can be obtained using the usual input–output principle [7].

$$T = \left| 1 - \frac{\sqrt{2k}a_+}{E_s} \right|, \quad (16.8)$$

and

$$a_{out+} = \sqrt{2k}a_+. \quad (16.9)$$

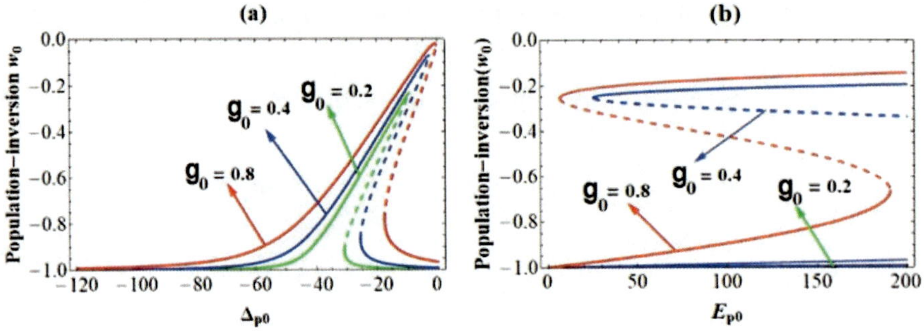

Fig. 16.2 **a** Population inversion w_0 w.r.t pump-exciton detuning Δ_{p0}; Here $E_{p0} = 64$, $\eta = 0.2$, $\omega_{\vec{k}} = 100$, $\Delta_{c0} = 10$, $k_{c0} = 1.35$. **b** The graph of Population inversion w_0 verses pump field. Parameters are used for graph are—$\Delta_{c0} = 10$, $\Delta_{p0} = -8$, $\eta = 0.2$, $\omega_{\vec{k}} = 100$, $k_{c0} = 1.35$. All these parameters are dimensionless w.r.t Γ_2

16.3 Results and Discussion

In this section we discuss the results of our system and effect of system parameters on the results.

16.3.1 Optical Bistability

Dynamic backaction is caused by radiation strain, due to which optical bistability occurs within the finite cavity decay time [8–16]. Here, we will address how bistability is affected by the strength of QD and cavity coupling (g_0).

In Fig. 16.2a, we map the population inversion w_0 with three separate values of g_0. Increasing the coupling of the QD-cavity mode results in more prominence of bistable behavior. Bistability disappears in the case of low values of g_0 and E_{p0}. Figure 16.2b demonstrates the usual bistable S-shaped curves for multiple values of g_0.

Bistability exists at a higher pump intensity value of E_{p0} and at a very low value of g_0. Bistability disappears with a decrease of g_0.

We can infer from the above discussion that a fine control over the QD-cavity mode coupling (g_0) is needed to tune the bistability that can be used to design all-optical switches, memory devices and logic-gate.

16.3.2 Controllable MIA and Fano Resonance

We demonstrate mechanically induced absorption (MIA) in our coupled QD-phonon system in this section. The mechanics appears in MIA because of lattice vibrations.

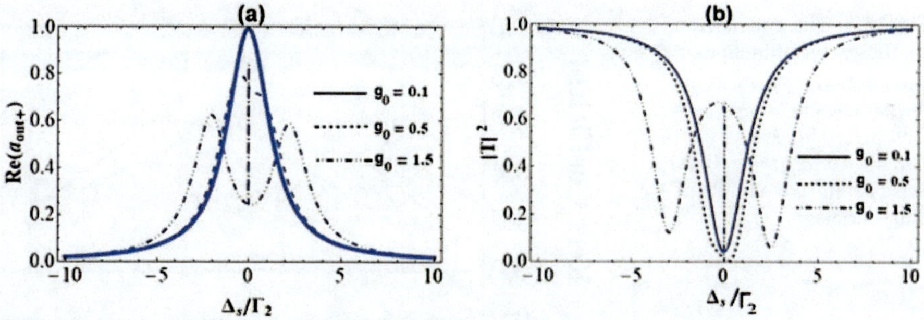

Fig. 16.3 **a** The graph of absorption a_{out+} w.r.t signal exciton detuning Δ_s for different value of coupling factor. **b** Graph of Transmission T^2 w.r.t signal exciton detuning Δ_s for different values of QD-cavity coupling; Parameters for graph are—$E_{p0} = 5$, $\eta = 0.015$, $\omega_{\overrightarrow{k0}} = 10$, $k_{c0} = 1.35$, $\Delta_{c0} = -10$, $\Delta_{p0} = -10$ (All these parameters are dimensionless w.r.t Γ_2)

The spectral profile of OMIT/OMIA is known to be symmetric while that of Fano-resonance is asymmetric. From Fig. 16.3a, we note that absorptive activity increases as the coupling strength between Q.D and the cavity decreases. Also as the $\boldsymbol{g}_0$ value is increased, the outline of the Fano line appears.

We use two optical fields for this problem, rather than a single field. Figure 16.3b displays the $|T|^2$ transmission as a function of Δ_s signal-exciton detuning. We can see that only a complete absorption dip (Solid line) is reflected in the transmission spectrum for small coupling between the cavity and quantum dot.

The transmission spectrum is splits into two peaks as g_0 rises gradually. We can see that as g_0 increases, the split width increases.

16.3.3 All Optical Kerr Switch

With the rapid advancement of optics, in order to make good use of it when it is wanted and to prevent it when it is unwanted, there is a realistic need to consider the nonlinear behavior of the optical material [17–20]. In different materials, the Kerr effect has drawn a great deal of interest. For the implementation of ultrafast switches, we use the instantaneous optical Kerr effect, which is responsible for self-phase modulation [21], modulation instability [22] and nonlinear effects of self-focusing [23]. Theoretical knowledge of an optomechanical Kerr switch based on lattice vibrations and a quantum dot is provided in this paper.

Using (16.7), If we set $\Delta_{p0} = 0$, (pump-exciton detuning) and $\Delta_{c0} = 0$ (pump cavity detuning), a different Kerr coefficient behavior can be seen in Fig. 16.4. The two peaks on either side of Δ_s=0 indicate the vibrational frequency of the phonon. This means that if we first set the control field of Δ_{p0} and Δ_{c0} and observe the signal frequency in the continuum around the exciton frequency ω_{ex}, then we can reliably get the vibrational frequency of the phonon within the nonlinear framework.

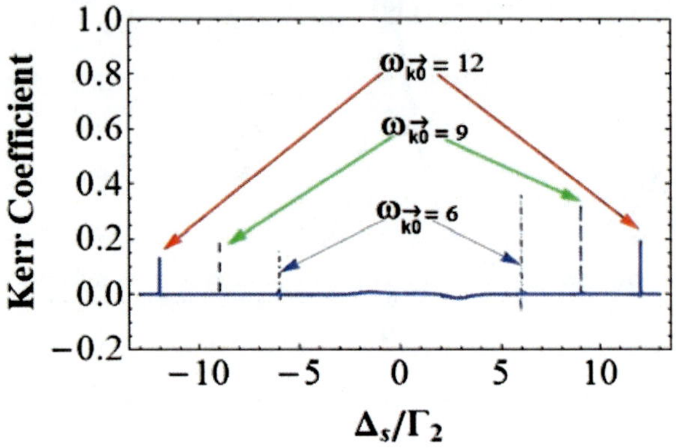

Fig. 16.4 The optical Kerr coefficient as a function of detuning Δ_s. For the graph, the parameters are—$E_{p0} = 0.34$, $\eta = 0.06$, $k_{c0} = 1.35$, $\Delta_{c0} = -10$, $\Delta_{p0} = -10$ (All these parameters are w.r.t Γ_2 dimensionless

16.4 Conclusion

We discussed the system's optical response containing a single quantum dot interacting with the optical mode of a solid state micro-cavity and vibrations of the lattice. We show that inherent nonlinearity in the system produces optical bistability which can be controlled by QD-cavity coupling. In particular, if the QD cavity mode coupling is strong, we demonstrate that the bistability occurs at a lower pump value. An asymmetric Fano resonance as a result of exciton-phonon coupling is visible, giving rise to anomalous dispersion in the transmitted signal, which means that it is possible to generate slow light using this system. Finally, we show the possibility of using the device in fast optical communication networks.

Acknowledgements P. K. Jha, A. Bhattacherjee and Vijay Bhatt are grateful to DST (SERB) New Delhi, for financial support under grant no. EMR/2017/001980. S. A. Barbhuiya acknowledges the Ph.D. fellowship of BITS, Pilani.

References

1. M. Aspelmeyer, T.J. Kippenberg, F. Marquardt, Rev. Mod. Phys. **86**, 1391 (2014)
2. T.J. Kippenberg, K.J. Vahala, Science, **321**, 1172 (2008)
3. F. Marquardt, S.M. Girvin, Physics **2**, 40 (2009)
4. M. Aspesmeyer, S. Groeblacher, K. Hammerer, N. Kielsel, J. Opt. Soc. Am. B **27**, A189 (2010)
5. J.J. Li, K.-D. Zhu, J. App. Phys. **110**, 114308 (2011)
6. J. Gudat, Cavity quantum electrodynamics with quantum dots in microcavities. Ph.D. Thesis, University of Leiden (2012)
7. D.F. Walls, G.J. Milburn, Quant. Opt.
8. M. Vengalattore, M. Hafezi, M.D. Lukin, M. Prentiss, Phys. Rev. Lett. **101**, 063901, (2008)
9. F. Brennecke, S. Ritter, T. Donner, T. Esslinger, Science **322**, 235 (2008)
10. T.P. Purdy, D.W.C. Brooks, T. Botter, N. Brahms, Z.Y. Ma, D.M. Stamper-Kurn, Phys. Rev. Lett. **105**, 133602 (2010)
11. H.A. Babu, H. Wanare, Phys. Rev. A **83**, 033818 (2011)

12. R. Ghobadi, A.R. Bahrampur, C. Simon, Phys. Rev. A **84**, 033846 (2011)
13. E.A. Sete, H. Eleuch, Phys. Rev. A **85**, 043824 (2012)
14. V. Bhatt, S.A. Barbuiya, P.K. Jha, A.B. Bhattacherjee, J. Phys. B: Mol. Opt. Phys. **53**, 155402 (2020)
15. M.K. Singh, P.K. Jha, A.B. Bhattacherjee, J. Mod. Opt. **67**(8), 692–703 (2020)
16. C. Jiang, H. Liu, Y. Cui, X. Li, G. Chen, X. Shuai, Phys. Rev. A **88**, 055801 (2013)
17. F.L. Semiao, K. Fueuya, G.J. Milburn, Phys. Rev. A **79**, 063811 (2009)
18. H. Hashemi, A.W. Rodriguez, J.D. Joannopoulos, M. Soljacic, S.G. Johnson, Phys. Rev. A **79**, 013812 (2009)
19. R. Lifshitz, M.C. Cross, Review of nonlinear dynamics and complexity, (Wiley) **1**, 52 (2008)
20. F.A. Malek, W. Aroua, S. Haxha, I. Flint, Annalen der Physik 528, (2016)
21. M.R.G. Robb, J.W.B. McNeil, Phys. Rev. Lett. **92**, 2 (2005)
22. B.C. Tabi, A. Mohamadou, T.C. Kofane, J. Phys.: Condens. Matter **21**, 33
23. S.M. Sodha, M. Faisal, P.M. Verma Phys. Plasmas **16**, 082304 (2009)

Chapter 17
Estimation of Electronic and Optical Properties of Chalcopyrite Semiconductors Using Machine Learning

S. K. Tripathy, J. K. Singh, G. M. Prasad, and F. A. Talukdar

Abstract In this paper, we have predicted the electronic and optical properties of $A^{I}B^{III}C_2{}^{VI}$ chalcopyrite semiconductors using Ada-Boost algorithm under the framework of machine learning. The values of bandgap (E_g), refractive index (n), bulk modulus (B) and dielectric constant (ε) of $A^{I}B^{III}C_2{}^{VI}$ chalcopyrite semiconductors are predicted employing lattice constants, bond length, ionicity (f_i) and plasmon energy ($\hbar\omega_p$) as input parameters by means of Ada-Boost algorithm. The results obtained were compared with available experimental and theoretical values. The result presented are found to be in good agreement with experimental values and having less average percentage deviation compared to reported theoretical values.

17.1 Introduction

The $A^{I}B^{III}C_2{}^{VI}$ and $A^{II}B^{IV}C_2{}^{V}$ compounds crystallize in the chalcopyrite structure with a tetrahedral space group $\left(I\bar{4}2d\right)$ having four formula units in each unit cell which are analogues to $A^{II}B^{VI}$ and $A^{III}B^{V}$ binary tetrahedral semiconductor, respectively. These chalcopyrite semiconductor has potential application in the fields of light emitting diodes (LEDs), laser diodes (LDs), photo detectors (PDs), photovoltaic cells, integrated optics, nano-electronics, frequency conversion and nonlinear optical (NLO) devices [1]. The dielectric constant of $A^{I}B^{III}C_2{}^{VI}$ and $A^{II}B^{IV}C_2{}^{V}$ chalcopyrite semiconductors have been studied versus bond length [2]. Further, electronic susceptibilities, energy gaps and the bond ionicities have been calculated using their plasmon energy [3]. Kumar et al. have proposed simple empirical relationship to

S. K. Tripathy (✉) · F. A. Talukdar
Department of Electronic and Communication Engineering, National Institute of Technology Silchar, Silchar, India
e-mail: sktripathy@ece.nits.ac.in

J. K. Singh · G. M. Prasad
Intelligent Mining Systems, CSIR-Central Institute of Mining and Fuel Research, Barwa Road, Dhanbad, India

R. G. Nair et al. (eds.), *Proceedings of 28th National Conference on Condensed Matter Physics*, Springer Proceedings in Physics 269,
https://doi.org/10.1007/978-981-16-5407-7_17

estimate bulk modulus (B) and microhardness (H) of group IV, II–VI, III–V, I–III–VI_2 and II–IV–V_2 semiconductors employing the plasma oscillations theory of solids [4]. Using Moss model [5] relationship between refractive index and bandgap energy can be established as $n = \left(\frac{K}{E_g}\right)^{1/4}$, where K is a constant (=95 eV) and E_g is bandgap energy. From the above discussion we have found that energy gap, bond ionicity, bulk modulus and microhardeness are related to plasmon energy directly or indirectly. Further, plasma energy can also be written as

$$\hbar\omega_p = 28.81\ (Z\,\sigma/W)^{1/2}\ \ eV$$

where Z is the number of valence electron taking part in the plasmon oscillations, σ is the specific gravity and W the molecular weight. Further, it is well known that Z is also related to lattice constant. It is also well known that bandgap (E_g), dielectric constants and refractive indices also depends on some of fundamental features of crystal such as lattice constants (a, c), bond length (d), plasmon energy $(\hbar\omega_p)$ and ionicity (f_i) [6, 7]. In this work, we have predicted the values of E_g, n and ε of $A^IB^{III}C_2{}^{VI}$ chalcopyrite semiconductors employing a, c, d, f_i and $\hbar\omega_p$ as input parameters by means of Ada-Boost algorithm. The results obtained were compared with available experimental and theoretical values. The result presented are found to be in good agreement with experimental values and having less average deviation compared to reported theoretical values (Table 17.1).

17.2 Methodologies

Machine learning is an artificial intelligence tool that enables any kind of system or hypothesis to learn from data instead of a pre-specified program. That means the outcome of machine learning assisted system will be the algorithm or program itself unlike the traditional system which gives any kind of output or data depending upon the input program. But the application of machine learning is not merely limited to computer science or technology. It possesses huge potential in any field where it can be applicable. Broadly machine learning is divided into two categories, supervised and unsupervised. In the work, adaptive boosting (Ada-Boost) algorithm is a method under supervised learning to achieve higher accuracy with combination of many weak and imprecise classifiers. It can utilize any kind of weak classifier and merge all the outcomes with the help of assigned weights to predict the final outcome [8]. Under this algorithm, the given training instances are (×1, y1), (xm, ym) where xi ∈ X, yi ∈ $\{-1, +1\}$. Several weak learner or weak learning algorithm is applied to find a weak hypothesis ht: X → $\{-1, +1\}$, where the purpose of that weak classifier is to look for a hypothesis with very less weighted error. The final or combined hypothesis F(x) computes the sign of a weighted combination of weak hypotheses

Table 17.1 Results of the prediction of refractive Index of I-III-VI_2 chalcopyrite semiconductor

Compounds	Lattice constant (Å)		Bond length (Å)	Ionicity	Plasmon energy (eV)	Refractive index						
						Expt	Reported				Cal	
1	2	3	4	5	6	7	8	9	10	11	12	13
	a [7]	*c* [7]	*d* [7]	f_i [12]	$\hbar\omega_p$ [4]	*n* [12–14]	[15]	[5]	[16]	*[17]*	*This work*	*PAD*
$CuAlS_2$	5.323	10.44	2.29	0.687	17.25	2.4	2.22	2.88	1.91	2.33	2.423	0.9583
$CuAlSe_2$	5.617	10.92	2.4	0.692	15.86	2.6	2.52	2.44	2.41	2.59	2.679	3.0384
$CuAlTe_2$	5.976	11	2.58	0.699	14.35	3.3	2.79	2.61	3.52	3.69	3.31	0.3030
$CuGaS_2$	5.36	10.49	2.3	0.689	17.1	2.67	2.63	2.51	2.53	2.71	2.594	2.8464
$CuGaSe_2$	5.618	11.01	2.416	0.696	15.92	2.8	2.77	2.74	3.03	3.06	2.692	3.8571
$CuGaTe_2$	6.013	11.92	2.602	0.698	14.3	3.3	3.31	3.12	3.46	3.59	3.315	0.4545
$CuInS_2$	5.528	11.08	2.4	0.689	16.12	2.6	3.03	2.82	3.15	3.18	2.6675	2.5961
$CuInSe_2$	5.785	11.56	2.509	0.695	15.09	2.9	3.16	3.15	3.49	3.63	2.775	4.3103
$CuInTe_2$	6.179	12.365	2.676	0.678	13.66	3.4	3.34	3.16	3.49	3.64	3.32	2.3529
$AgAlS_2$	5.695	10.26	2.4	0.704	16.26	2.3	2.35	2.35	2.15	2.45	2.4	4.3478
$AgAlSe_2$	5.968	10.77	2.51	0.708	15.16	2.47	2.58	2.47	2.5	2.66	2.565	3.8461
$AgAlTe_2$	6.309	11.85	2.68	0.713	14.5	2.54	2.7	2.55	2.68	2.77	2.6417	4.0039
$AgGaS_2$	5.753	10.28	2.42	0.724	16.1	2.4	2.51	2.44	2.41	2.59	2.46	2.5
$AgGaSe_2$	5.985	10.9	2.525	0.709	14.76	2.8	2.9	2.7	2.96	3	2.695	3.75
$AgGaTe_2$	6.301	11.96	2.69	0.713	13.63	3.3	3.2	3.05	3.4	3.49	3.316	0.4848
$AgInS_2$	5.828	11.19	2.49	0.708	15.21	2.5	2.88	2.63	2.84	2.89	2.604	4.16
$AgInSe_2$	6.102	11.69	2.61	0.712	14.23	3.32	3.19	2.96	3.31		3.313	0.2108
$AgInTe_2$	6.42	12.59	2.773	0.716	13.04	3.4	3.3	3.12	3.46	3.66	3.3275	2.1323

(continued)

Table 17.1 (continued)

Compounds	Lattice constant (Å)		Bond length (Å)	Ionicity	Plasmon energy (eV)	Refractive index						
						Expt	Reported				Cal	
1	2	3	4	5	6	7	8	9	10	11	12	13
$CuTlS_2$	5.58	11.17	2.42	0.6324	15.89						2.95	
$CuTlSe_2$	5.844	11.65	2.53	0.6335	14.88						2.806	
$CuFeS_2$	5.289	10.423	2.297	0.6351	17.19		3.27				2.819	
Average % deviation							**5.68**	**6.91**	**7.49**	**9.21**	**2.564**	

$$H = F(x) = \sum_{t=1}^{T} \alpha_t h_t(x)$$

where F(x) is weighted majority vote of the several weak hypotheses ht and αt is the weight for corresponding ht.

17.3 Results and Discussion

The refractive index of any material is defined as the ratio of velocity of light in air to the velocity of light in that material. From the literature survey, we found out that refractive index of a material is known to decrease with energy gap and therefore, these two fundamental quantities of a material are believed to have certain correlation. However, we should know that precise calculation of energy gap is another constraint in developing a model for refractive index. Therefore, we have employed fundamental attributes such as lattice constants, bond length, ionicity, plasmon energy instead of energy gaps to propose the machine learning assisted model. For refractive index, the maximum average deviation is 4.347% for $AgAlS_2$ and the average deviation is 2.564% for I-III-VI_2 structure. The average deviation is very less as compared to reported values by other authors. This is because the dataset for refractive index has a higher number of instances as compared to other parameters and the variation in the experimental values are also very less. All the experimental values are in the vicinity of 2.3 to 3.7. That should be the reason for less error shown by the model in the prediction of refractive index of chalcopyrite semiconductor.

Bulk modulus is the measure of compressibility of any material defined as the ratio of small change in pressure or volumetric stress to volumetric strain. Further, Verma et al. [9] has proposed a very simple relation between bulk modulus and plasmon energy. In this paper, we have used Ada-Boost algorithm to best fit the input parameters with bulk modulus. For bulk modulus the maximum deviation is 15.96% for $AgInTe_2$. Our average deviation for bulk modulus is limited to 3.477% for I-III-VI_2 structure and is very less compared to other earlier proposed model. The 3.477% of average deviation may be due to high standard deviation and also because of less number of experimental values are available. Also, it is found that compounds with less value of bulk modulus are more prone to error compared to high value of bulk modulus.

Bandgap energy is the distance between the valence and conduction band of material in eV or the minimum amount of energy required to migrate a single electron from valence to conduction band. While dielectric constant is defined as the ratio of permittivity of the material to the permittivity of free space. Philips [10] has resolved the bandgap energy into the homopolar and heteropolar nature of energies and dielectric constant can be expressed $\varepsilon = 1 + \left(\frac{\omega_p}{E_g}\right)^2$. Kumar et al. [11] proposed a simple relation for bandgap energy and dielectric constant in terms of plasmon energy for II-VI and III-V binary semiconductor. From the above explanation it is clear that bandgap energy and dielectric constant is related to bond length. Since bond length is the function of plasmon energy, we can consider that bandgap energy and dielectric constant is the function of bond length, lattice constant and plasmon energy. We have used such basic features to build a model for bandgap energy and dielectric constant using machine learning. The values calculated from this model has maximum deviation of 23.214% and 8.8847% for $AgAlTe_2$ and $AgAlS_2$, respectively. This deviation is pretty high as compared to any parameters predicted by this model. However, the average % deviation is only 7.8307 and 5.055, respectively, for bandgap energy and dielectric constants predicted for 22 ternary semiconductors. Also, no reported values are available to compare with our proposed model (Table 17.2).

17.4 Conclusion

In this work, we have predicted the four properties such as refractive index, bulk modulus, bandgap energy and dielectric constant of I-III-VI_2 group of chalcopyrite semiconductors. This approach using a machine learning algorithm for the prediction of properties works well and has shown improve average percentage deviation over previous models, which are based on expression. We have also predicted the values of all the parameters of $CuTlS_2$, $CuTlSe_2$ and $CuFeS_2$ of I-III-VI_2 group. These compounds have been very less studied by the previous authors. However, unavailability of the experimental data for these compounds restricts us to show the percentage error which needs further investigation (Table 17.3).

Table 17.2 Results of the prediction of bulk modulus Index of I-III-VI_2 chalcopyrite semiconductor

Compounds	Lattice constant (Å)		Bond length (Å)	Ionicity	Plasmon energy (eV)	Bulk modulus (GPa)						
						Expt	Reported				Cal	
1	2	3	4	5	6	7	8	9	10	11	12	13
	a [7]	c [7]	d [7]	f_i [12]	$\hbar\omega_p$ [4]	B [18]	[4]	[19]	[20]	[21]	*This work*	*PAD*
$CuAlS_2$	5.323	10.44	2.29	0.687	17.25	94[a], 99[b]	84.02	93	96	85.3	93.6	0.4255
$CuAlSe_2$	5.617	10.92	2.4	0.692	15.86	84[b]	69.06	68	80	72.9	80.028	4.7285
$CuAlTe_2$	5.976	11	2.58	0.699	14.35		54.68	53	64	53.5	54.33	
$CuGaS_2$	5.36	10.49	2.3	0.689	17.1	94[a], 96[j] 97[d]	82.32	91	94	98	93.55	0.4787
$CuGaSe_2$	5.618	11.01	2.416	0.696	15.92	71[c], 94[d]	69.67	70	79	69.3	76.5	7.7464
$CuGaTe_2$	6.013	11.92	2.602	0.698	14.3	44[c]	54.24	51	38	48.5	44	0
$CuInS_2$	5.528	11.08	2.4	0.689	16.12	71.1	71.73	70	81	72.9	77.94	3.92
$CuInSe_2$	5.785	11.56	2.509	0.695	15.09	53.6, 48 [k] 62 [m], 72[e]	61.49	61	69	64	54.267	2.3905
$CuInTe_2$	6.179	12.365	2.676	0.678	13.66	36[f], 45.4 [m]	48.74	45	51	45	36.857	2.3805
$AgAlS_2$	5.695	10.26	2.4	0.704	16.26		73.19		82	75.2	92.367	
$AgAlSe_2$	5.968	10.77	2.51	0.708	15.16		62.15	59	70	62.8	54.8	
$AgAlTe_2$	6.309	11.85	2.68	0.713	14.5		56.02	43	54	40.4	42	
$AgGaS_2$	5.753	10.28	2.42	0.724	16.1	90 [g], 60[j] 67[n], 72 [h]	71.52	70	76	67.2	91.025	1.1388
$AgGaSe_2$	5.985	10.9	2.525	0.709	14.76	54.8 59.8 [h], 65 [g]	58.39	55	63	59.9	54.625	0.3193
$AgGaTe_2$	6.301	11.96	2.69	0.713	13.63	35.7[f], 6.6 [h] 71.5[p]	48.49	42	21	38.9	38.85	8.8235

(continued)

Table 17.2 (continued)

Compounds	Lattice constant (Å)		Bond length (Å)	Ionicity	Plasmon energy (eV)	Bulk modulus (GPa)						
						Expt	Reported				Cal	
1	2	3	4	5	6	7	8	9	10	11	12	13
$AgInS_2$	5.828	11.19	2.49	0.708	15.21	54.1	62.63	57	66	55	54.3	0.3696
$AgInSe_2$	6.102	11.69	2.61	0.712	14.23	42	53.62	47	50	48.7	42	0
$AgInTe_2$	6.42	12.59	2.773	0.716	13.04	29.6	43.73	38	29	24.5	34.325	15.962
$CuTlS_2$	5.58	11.17	2.42	0.6324	15.89						79.6	
$CuTlSe_2$	5.844	11.65	2.53	0.6335	14.88						54.2	
$CuFeS_2$	5.289	10.423	2.297	0.6351	17.19						94	
Average % deviation							11.77	10.1	10.42	5.69	3.477	

[a][22], [b][23], [c][24], [d][25], [e][26], [f][27], [g][28], [h][29], [j][30]

Table 17.3 Results of the prediction of bandgap energy and dielectric constant of I-III-VI_2 chalcopyrite semiconductor

Compounds	Lattice constant (Å)		Bond length (Å)	Ionicity	Plasmon energy (eV)	Bandgap energy (eV)				Dielectric constant		
						Expt	Cal		Expt	Reported	Cal	
1	2	3	4	5	6	7	8	9	10	11	12	13
	a [7]	*c* [7]	*d* [7]	f_i [12]	$\hbar\omega_p$ [4]	e_g [1]	*This work*	*PAD*	ε [12–14]	[31]	*This work*	*PAD*
$CuAlS_2$	5.323	10.44	2.29	0.687	17.25	$3.49^{\&}$	3.5	0	5.76	5.945	6.0483	5.005
$CuAlSe_2$	5.617	10.92	2.4	0.692	15.86	$2.71^{\&}$	1.676	0.2380	6.76	6.584	7.2554	7.3284
$CuAlTe_2$	5.976	11	2.58	0.699	14.35	0.88*	0.9128	3.7272	10.89	7.544	10.89	0
$CuGaS_2$	5.36	10.49	2.3	0.689	17.1	2.43	2.472	1.7283	7.1289	6.226	6.994	1.8923
$CuGaSe_2$	5.618	11.01	2.416	0.696	15.92	1.68	1.665	0.8928	7.84	6.827	7.3986	5.6301
$CuGaTe_2$	6.013	11.92	2.602	0.698	14.3	1.0	0.84875	15.125	10.89	7.664	10.9694	0.7291
$CuInS_2$	5.528	11.08	2.4	0.689	16.12	1.53	1.659	8.4313	6.76	6.943	7.2554	7.3284
$CuInSe_2$	5.785	11.56	2.509	0.695	15.09	1.04	1.09125	4.9278	8.41	7.085	7.75	7.8478
$CuInTe_2$	6.179	12.365	2.676	0.678	13.66	1.06	0.963	9.1509	11.56	8.153	11.0224	4.6505
$AgAlS_2$	5.695	10.26	2.4	0.704	16.26		2.5		5.29	6.513	5.76	8.8847
$AgAlSe_2$	5.968	10.77	2.51	0.708	15.16	0.7*	0.7167	2.3857	6.1009	7.124	6.5669	7.6382
$AgAlTe_2$	6.309	11.85	2.68	0.713	14.5	0.56*	0.69	23.2142	6.4516	8.123	6.951	7.7407
$AgGaS_2$	5.753	10.28	2.42	0.724	16.1	2.64	2.57	2.65151	5.76	6.623	6.055	5.1215
$AgGaSe_2$	5.985	10.9	2.525	0.709	14.76	1.8	1.613	10.3888	7.84	7.209	7.24	7.6531
$AgGaTe_2$	6.301	11.96	2.69	0.713	13.63	1.32	1.155	12.5	10.89	8.156	10.996	0.9734
$AgInS_2$	5.828	11.19	2.49	0.708	15.21	1.87	1.6587	11.2994	6.25	7.026	6.7258	7.6128
$AgInSe_2$	6.102	11.69	2.61	0.712	14.23	1.24	1.09625	11.5927	11.0224	7.689	10.9893	0.3003
$AgInTe_2$	6.42	12.59	2.773	0.716	13.04	0.95	1.09125	14.8684	11.56	8.729	11.0224	4.6505

(continued)

Table 17.3 (continued)

Compounds	Lattice constant (Å)		Bond length (Å)	Ionicity	Plasmon energy (eV)	Bandgap energy (eV)				Dielectric constant		
						Expt	Cal		Expt	Reported	Cal	
1	2	3	4	5	6	7	8	9	10	11	12	13
$CuTlS_2$	5.58	11.17	2.42	0.6324	15.89		1.6525				7.8936	
$CuTlSe_2$	5.844	11.65	2.53	0.6335	14.88		0.964				8.2675	
$CuFeS_2$	5.289	10.423	2.297	0.6351	17.19		3.5				7.1289	
Average % deviation							**7.8307**			**17.82**	**5.055**	

&[32], *33]

Acknowledgements The authors are thankful to CSIR, New Delhi, Government of India for grant under Extra Mural Research-II scheme (File No. 70(0076)/19/EMR-II). We are also thankful to Prof. Sivaji Bandyopadhyay, Director, National Institute of Technology, Silchar for his continuous encouragement and inspiration in conducting this work.

References

1. J.L. Shay, J.H. Wernick, *Ternary Chalcopyrite Semiconductors: Growth, Electronic Properties, and Applications* (Pergamon Press, New York, 1975)
2. A.S. Verma, D. Sharma, Phys. Scr. **76**(1), 22 (2007)
3. V. Kuṁar, J. Phys. Chem. Solids **48**(9), 827 (1987)
4. V. Kumar, A.K. Shrivastava, V. Jha, J. Phys. Chem. Solids **71**(11), 1513 (2010)
5. T.S. Moss, Physica status solidi (b) 131(2), 415 (1985)
6. V. Kumar, J.K. Singh, Ind. J. Pure Appl. Phys. **48**, 571 (2010)
7. V. Kumar, B.S.R. Sastry, J. Phys. Chem. Solids **63**, 107 (2002)
8. Y. Freund, R.E. Schapire, J. Comput. Syst. Sci. **55**(1), 119 (1997)
9. A.S. Verma, S.R. Bhardwaj, Phys. Status Solidi B **243**(12), 2858 (2006)
10. J.C. Philips, Rev. Mod. Phys. **42**, 317 (1970)
11. V. Kumar, A.K. Shrivastava, A. Sinha, V. Jha, Ind. J. Pure Appl. Phys. **51**, 49 (2013)
12. V.P. Gupta, N.M. Ravindra, Phys. Status Solidi B **100**, 715 (1980)
13. T.S. Moss, Proc. Phys. Soc. B **63**, 167 (1950)
14. V. Gopal, Infrared Phys. **22**, 255 (1982)
15. V. Kumar, A. Sinha, B.P. Singh, A.P. Singh, V. Jha, Chin. Phys. Lett. **32**(12), 127701–127711 (2015)
16. N.M. Ravindra, S. Auluck, V.K. Srivastava, Phys. Status Solidi B **93**, K155 (1979)
17. R.R. Reddy, S. Anjaneyulu, Phys. Status Solidi B **174**, K91 (1992)
18. H. Neumann, Crystal Res. Technol. 23, 97 (1988) Phys. Status Solidi 96, K121 (1986)
19. A.S. Verma, Mater. Chem. Phys. **139**(1), 256 (2013)
20. P.G. Gallardo, Phys. Status Solidi B **182**, K67 (1994)
21. S.H. Wei, A. Zunger, I.H. Choi, P.Y. Yu, Phys. Rev. B **58**, R1710 (1998)
22. M. Bettini, W.B. Holzapfel, Solid State Commun. **16**, 27 (1975)
23. L. Roa, J.C. Chervin, J.P. Itie, A. Polian, M. Gauthier, A. Chevy, Phys. Status Solidi B **211**, 455 (1999)
24. A. Kraft, G. Kuhn, W. Moller, Z. anorg. allg. Chem. **504**, 155 (1983)
25. T. Tinoco, J.P. Itie, A. Polian, A. San, E.M. Miguel, P. Grima, J. Gonzalez, F. Gonzalez, J. Phys. **4**(C9), 151 (1994)
26. T. Tinoco, A. Polian, D. Gomez, J.P. Itie, Phys. Status Solidi B **198**, 433 (1996)
27. D.R. Lide, Hand book of Chemistry and Physics, 80th ed., CRC Publication, 1999–2000
28. T. Tinoco, A. Polian, J.P. Itie, E. Moya, J. Gonzalez, J. Phys. Chem. Solids **56**, 481 (1995)
29. Y. Mori, K. Takarabe, S. Iwamoto, S. Minomura, E. Niwa, K. Masumoto, Phys. Status Solidi B **198**, 427 (1996)
30. A. Werner, H.D. Hochheimer, A. Jayaraman, Phys. Rev. B **23**, 3836 (1981)
31. A.S. Verma, D. Sharma, Phys. Scr. **76**, 22 (2007)
32. J.E. Jaffe, A. Zunger, Phys. Rev. **28**(10), 5822 (1983)
33. M. Haynes William, Handbook of Chemistry and Physics, 92nd ed., CRC Press, Boca Raton (2011)

Chapter 18
Flow of Medium Constituent with Charged Magnetic Particles in Presence of External Magnetic Field

Srikanta Debata, Tanmay Das, Jayanta Dey, Dhruv Pratap Singh, Sesha Vempati, and Sabyasachi Ghosh

Abstract This study explores the dissipation flow of Newtonian fluid in the presence of finite rotation, and the external magnetic field. We have considered a simple non-interacting scenario of fluid medium under the relaxation time approximation of kinetic theory framework. We have identified the equivalence roles of Lorentz force and Coriolis force in finite magnetic field and rotation, respectively. This has been shown by detailed microscopic calculations of electrical conductivity. We have found an anisotropic conductivity tensor due to rotation as normally observed in the system with finite magnetic field.

18.1 Introduction

Transport coefficients like electrical conductivity, viscosity etc. can be microscopically calculated from relaxation time approximation (RTA) of kinetic theory approach [1, 2]. A simplest RTA-based expression of conductivity as proposed by Drude in 1900 [3] explains conductivity of metals via free electron theory. The relativistic counterpart of RTA (or Field theoretical Kubo [4] expression) for nuclear matter can

S. Debata (✉) · T. Das · J. Dey · D. P. Singh · S. Vempati · S. Ghosh
Indian Institute of Technology Bhilai, GEC Campus, Sejbahar, Raipur, Chhattisgarh 492015, India
e-mail: srikantad@iitbhilai.ac.in

T. Das
e-mail: tanmayd@iitbhilai.ac.in

J. Dey
e-mail: jayantad@iitbhilai.ac.in

D. P. Singh
e-mail: dhruv@iitbhilai.ac.in

S. Vempati
e-mail: sesha@iitbhilai.ac.in

S. Ghosh
e-mail: sabya@iitbhilai.ac.in

R. G. Nair et al. (eds.), *Proceedings of 28th National Conference on Condensed Matter Physics*, Springer Proceedings in Physics 269,
https://doi.org/10.1007/978-981-16-5407-7_18

be found in [5] and references therein. Notably, the RTA easily handles dissipation and it is frequently applied on electronic and thermal properties of graphene [6] and cold atom [7] systems. Such many body systems in the presence of an external magnetic field and medium's vorticity (rotation) might depict rather interesting phenomena. One of such phenomena is Barnett effect (1917), in which rotating a ferromagnetic object at frequency $\vec{\omega}$ in the absence of a magnetic field produces the same magnetization as when the object is not rotating but in the presence of a magnetic field $\vec{B}_{\text{eff}} = \frac{\vec{\omega}}{\gamma}$, where γ is the gyromagnetic ratio [8–10]. Years later the Barnett effect was proved experimentally [11, 12] for example, Arabgol et al. shows that a spinning of water at very high frequencies leads to the magnetic polarization [12]. The other interesting phenomena are chiral vortical effect [13], chiral magnetic effect [14, 15] and local spin polarization [16, 17].

In this background, here we attempt to understand how the microscopic properties like electric charge and magnetic moment of the individual particles influences the transport properties of many body systems in different external conditions like magnetic field, vorticity. We follow a simplistic and rather powerful approach based on the RTA of the kinetic theory framework.

The article is organized as follows. Formalism explores microscopic calculation of electrical conductivity, which changes in presence of rotation and external magnetic field. For the case of rotation, the angular frequency of the medium can couple with velocity of medium constituents (Coriolis force). In the case of magnetic fields the magnetic moments of medium constituents can couple with the external magnetic field (Larmor precession), which we are currently pursuing. A graphical representation is provided in the result section to show the impact of rotation on electrical conductivity and at the end, we have summarized our findings.

18.2 Formalism

RTA framework assumes a small deviation δf from the equilibrium distribution function of constituents within a time interval τ_c, called relaxation time. Electrical current density in kinetic theory and Ohm's law can be written as $(J_x) = ge^2 \int \frac{d^3\vec{k}}{(2\pi)^3} \frac{k_x^2}{m^2} \delta f = (\sigma_{xx})(E_x)$, where electric field E_x is applied along x axis and medium constituent particles carry degeneracy factors g, electric charge e, mass m, energy $\epsilon = \frac{|\vec{k}|^2}{2m}$. With the help of RTA-based Botzman equation, one can express δf in terms of τ_c and equilibrium distribution function f_0 [18]. This yields an expression of conductivity along the x-direction [18].

$$\sigma_{xx} = ge^2\beta \int \frac{d^3\vec{k}}{(2\pi)^3} \tau_c \frac{|\vec{k}|^2}{3m^2} f_0(1 \mp f_0) \tag{18.1}$$

Depending upon Fermi–Dirac (FD) or Bose–Einstein (BE) distribution function f_0 at temperature $T = 1/\beta$, the minus and plus signs will come in the phase-space factor. In general, the electrical conductivity tensor of electron gas is isotropic in nature, i.e.

$$\sigma_{xx} = \sigma_{yy} = \sigma_{zz}$$

This property may not be isotropic for rotating charged fluid/gas. The general procedure for obtaining the isotropic conductivity for non-rotating gas is almost the same for anisotropic conductivity of rotating gas. Here, only an additional term accounting the force creates the transition from isotropic to anisotropic picture. Therefore, we skip the discussion of isotropic conductivity and directly go into the details of anisotropic counterpart of rotating charged gas using RTA.

The rotating fluid with angular velocity $\overrightarrow{\omega}$ and particle with energy ϵ, the force term $\vec{F} \cdot \vec{\nabla}_k f$ in RTA of Boltzmann equation (BE) [18] will be

$$\left[e\vec{E} + 2m(\vec{v} \times \vec{\omega})\right] \cdot \vec{v}\, \frac{\partial f}{\partial \epsilon} = -\frac{\delta f}{\tau_c} \tag{18.2}$$

First term is for the electric field and second is the Coriolis force. For rotating gas, the second term is newly added and eventually, we will see that it will be responsible to create anisotropy in the medium. RTA based Boltzmann transport equation basically assumes non-equilibrium distribution $f = f_0 + \delta f$ carry a very small deviation δf from its equilibrium distribution function f_0 [18]. So one can put $f = f_0$ in the LHS of BE but the second term vanishes due to the vector identity $(\vec{v} \times \vec{\omega}) \cdot \vec{v} = \vec{\omega} \cdot (\vec{v} \times \vec{v}) = 0$. Therefore to include a rotation-dependent term, we consider the $\vec{\nabla}_k(\delta f)$ term also in BE,

$$e\vec{E} \cdot \vec{v}\, \frac{\partial f_o}{\partial \epsilon} + 2m(\vec{v} \times \vec{\omega}) \cdot \vec{\nabla}_k(\delta f) = -\frac{\delta f}{\tau_c} \tag{18.3}$$

where we take an ansatz

$$\begin{aligned} \delta f &= -(\vec{k} \cdot \vec{F})\frac{\partial f_0}{\partial \epsilon} \\ &= -(\vec{k} \cdot \vec{F})\beta f_0(1 - f_0) \end{aligned} \tag{18.4}$$

The unknown effective force by considering the resultant directions of electric field and Coriolis force will be

$$\vec{F} = (A_x\hat{x} + A_z\hat{z} + A_y(\hat{x} \times \hat{z})) \tag{18.5}$$

Reader can understand that for non-rotating case, $\vec{F} = A_x\hat{x} =\propto E_x\hat{x}$ but for rotating case, two more components appear along the direction of angular frequency $\hat{z}$, and coriolis force $\hat{x} \times \hat{z}$. Our goal is to find the coefficients A_x, A_y and A_z associated with the three directions. Now, using (18.4), (18.5) in (18.3), we get

$$\vec{v}.\left(e\vec{E}-2m\left(\vec{\omega}\times\vec{F}\right)\right)=\frac{\vec{k}.\vec{F}}{\tau_c},\tag{18.6}$$

where we have used a standard vector identity,

$$\begin{aligned}2m(\vec{v}\times\vec{\omega})\cdot\vec{\nabla}_k\delta f &= 2m(\vec{v}\times\vec{\omega})\cdot\nabla_k(-\vec{k}\cdot\vec{F})\frac{\partial f_0}{\partial\epsilon}\\ &= -2m(\vec{v}\times\vec{\omega})\cdot\vec{F}\,\frac{\partial f_0}{\partial\epsilon}\\ &= -2m\vec{v}\cdot(\vec{\omega}\times\vec{F})\frac{\partial f_0}{\partial\epsilon}\end{aligned}\tag{18.7}$$

For non-relativistic dispersion relation $\epsilon=\frac{|\vec{k}|^2}{2m}$,

$$\left(\frac{\vec{k}}{m}\right)\cdot[\mathrm{e}\vec{E}-2m(\vec{\omega}\times\vec{F})]=\frac{\vec{k}\cdot\vec{F}}{\tau_c}\tag{18.8}$$

The coefficients of $\hat{x}$, $\hat{y}$, *and* $\hat{z}$ in (18.6) will give us relation

$$A_z=0\tag{18.9}$$

$$A_x=\frac{\frac{e\tau_c}{m}}{1+4{\tau_c}^2\omega}E_x\tag{18.10}$$

$$A_y=\frac{\frac{-2{\tau_c}^2e\omega}{m}}{1+4{\tau_c}^2\omega}E_x\tag{18.11}$$

Hence, (18.4) can be written as

$$\delta f=e\tau_c\left(\frac{k_x}{m}-\frac{2k_y\tau_c\omega}{m}\right)E_x\frac{1}{1+4{\tau_c}^2\omega}\beta f_0(1-f_0)\tag{18.12}$$

Equation (18.12) indicates two directional current density based on Ohm's law,

$$\begin{aligned}(J_x)&=ge^2\int\frac{d^3\vec{k}}{(2\pi)^3}\frac{k_x^2}{m^2}\delta f=(\sigma_{xx})(E_x),\\ (J_y)&=ge^2\int\frac{d^3\vec{k}}{(2\pi)^3}\frac{k_y^2}{m^2}\delta f=(\sigma_{xy})(E_x)\end{aligned}\tag{18.13}$$

where,

$$\sigma_{xx} = ge^2\beta \int \frac{d^3\vec{k}}{(2\pi)^3}\tau_c \frac{1}{1+4\tau_c{}^2\omega}\frac{k_x{}^2}{m^2} f_0(1 \mp f_0) \tag{18.14}$$

$$\sigma_{yx} = -ge^2\beta \int \frac{d^3\vec{k}}{(2\pi)^3}\tau_c \frac{2\tau_c\omega}{1+4\tau_c{}^2\omega}\frac{k_y{}^2}{m^2} f_0(1 \mp f_0) \tag{18.15}$$

By applying $\vec{E} = E_y\hat{y}$, and repeating the same calculation yields expressions for σ_{yy}, σ_{xy}, which follow the relations $\sigma_{xx} = \sigma_{yy}, \sigma_{xy} = -\sigma_{yx}$. We can call them as perpendicular and Hall conductivity, respectively of the rotating fluid/gas. Similarly, longitudinal conductivity along z-axis will remain unaffected by rotation as Coriolis force does not work in that direction and its expression will be

$$\sigma_{zz} = ge^2\beta \int \frac{d^3\vec{k}}{(2\pi)^3}\tau_c \frac{k_z{}^2}{3m^2} f_0(1 \mp f_0) \tag{18.16}$$

Now let us move to another aspect of the medium in the presence of an external magnetic field, for which the magnetic moment of the medium constituents will execute a rotation around the field axis, commonly known as Larmor Precession. Let us assume an external magnetic field $\vec{B} = B\hat{z}$ is applied to the gas of charged particles with an angular momentum $\vec{L}$. Hence, the force term $\frac{\partial \vec{k}}{\partial t} \cdot \frac{\partial f}{\partial \vec{k}}$ in RTA of BE will be replaced as

$$\frac{\partial \vec{k}}{\partial t} \cdot \frac{\partial f}{\partial \vec{k}} + \frac{\partial \vec{L}}{\partial t} \cdot \frac{\partial f}{\partial \vec{L}} = -\frac{\delta f}{\tau_c}$$

$$\left(e\vec{E}.\vec{v} - \gamma\vec{B} \times \vec{L} \cdot \vec{\omega}\right)\frac{\partial f_0}{\partial \epsilon} = -\frac{\delta f}{\tau_c} \tag{18.17}$$

where $\gamma = \frac{e}{2m}$. The second term of the LHS is B-dependent and will again (like earlier rotation case) vanish due to the vector identity $-\left(\gamma\vec{B} \times \vec{L}\right)\cdot\vec{\omega} = \gamma\vec{B}.\left(\vec{L} \times \vec{\omega}\right) = 0$. So, to include the effect of, we consider the $\vec{\nabla}_L(\delta f)$ term also in BE.

$$e\vec{E} \cdot \vec{v}\,\frac{\partial f_0}{\partial E} - \gamma\vec{B} \times \vec{L} \cdot \vec{\nabla}_L(\delta f) = -\frac{\delta f}{\tau_c} \tag{18.18}$$

Here, particle movement and Larmor's precession are not directly connected in non-interacting gas, therefore the conductivity expression will not be affected by this mechanism. However, for interacting fluid/gas some changes are possible. Detailed study to investigate this problem is in progress. Also, in the presence of magnetic field, expressions for conductivity can be obtained as in the case of rotating fluid by replacing the Coriolis force with the Lorentz force [18].

18.3 Results and Discussion

The final expression of conductivity components for rotating fluid or gas in (18.14) and (18.15) are plotted with reference to the angular velocity ω in Fig. 18.1. We have applied an electrical field along x-direction to a rotating fluid, whose angular velocity is along the z-axis. Now, due to Coriolis force along y-direction, the medium constituents will be deviated from their previous electric field driven motion. They now face both forces—electric field along x-direction and Coriolis force along y-direction. In the RTA picture, it is relaxation time, which controls the strength of current density along x-direction. More explicitly, it fixes the magnitude of conductivity, which is understood as an intrinsic property of the medium. This relaxation time carries information of the microscopic collision within the medium. Hence a gross proportional relation between relaxation time and electrical conductivity is expected in RTA calculation for non-rotating fluid/gas. Now, when we switch on the rotation in the medium, then the y-directional current density can be expected due to Coriolis force and one can expect a Hall-type conductivity component (similar to the context to the presence of finite magnetic field). In fact Hall conduction at finite magnetic fields and finite rotation are exactly equivalent. Only the role of Lorentz force under finite magnetic field is replaced by Coriolis force under finite rotation. From Fig. 18.1, we notice that the conductivity along x-direction is reduced due to rotation and its reduction is increased as we increase the values of angular frequency (for a fixed value of relaxation time of a particular system). One can think an effective relaxation time $\tau_c\left(\frac{1}{1+4\tau_c{}^2\omega}\right)$, which decreases from its original value τ_c as the rotational motion of the fluid increases. On the other hand, Hall-type component

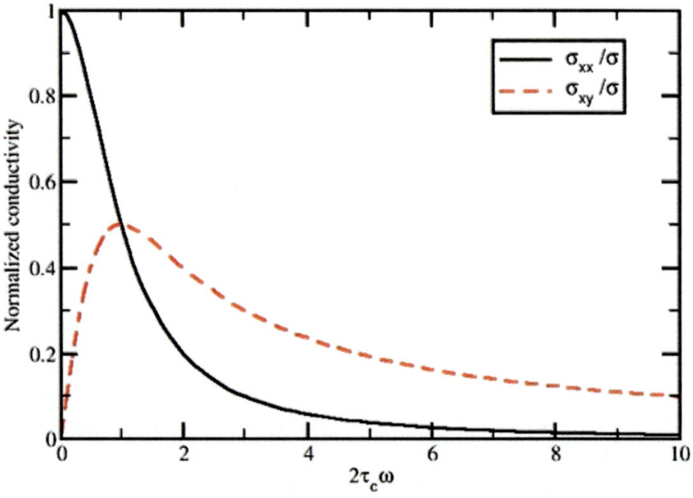

Fig. 18.1 Normalized electrical conductivity along x-direction (in direction of electric field) and xy-direction (like Hall component) are plotted against a dimensionless quantity $2\tau_c\omega$

is considered with effective relaxation time $\tau_c\left(\frac{2\tau_c\omega}{1+4\tau_c{}^2\omega}\right)$, which completely disappeared in the non-rotating medium. Interestingly this rotation-induced conductivity first increases then decreases with angular frequency. For a particular medium with fixed relaxation time, when the time period of rotation becomes twice the relaxation time value, this Hall-type conductivity will show its peak value. From Fig. 18.1, we get the clear picture of vorticity-dependent conductivity components (perpendicular and Hall). Similar kind of feature is also noticed in Hall conductivity at finite magnetic field [18]. In that case, when the inverse of cyclotron frequency or cyclotron time period become exactly the same with relaxation time, one can get peak value of Hall conductivity [18].

18.4 Conclusion and Outlook

In conclusion, our study shows an equivalent role between Lorentz and Coriolis forces in the presence of a finite magnetic field and rotation picture, respectively where electrical conduction or flow become anisotropic in nature. This scenario can be further extended by including interaction among the constituents as well as the effect of Larmor precession with finite magnetic moments. The work is under progress and aimed to apply in a different realistic system, which might reveal some noticeable response to finite rotation and magnetic field.

Acknowledgements Authors S.D., T.D., and J.D., acknowledge MoE for financial support.

References

1. C. Kittle, *Introduction to Solid State Physics*, 5th edn. (Wiley Eastern Limited, New Delhi, India, 1993)
2. E.M. Lifschitz, L.P. Pitajewski, *Physical Kinetics*, Landau series vol. 10, (Elsevier publisher, New Delhi, India, 2005)
3. P. Drude, Annalen der Physik 306, 566 (1900); 308, 369 (1900)
4. R. Kubo, J. Phys. Soc. Jpn. **12**, 570 (1957)
5. S Ghosh, Phys. Rev. D 95, 036018 (2017)
6. A.H. Castro Neto, F. Guinea, N.M.R. Peres, K.S. Novoselov, A.K. Geim, Rev. Mod. Phys. **81**, 109 (2009)
7. Y. Kikuchi, K. Tsumura, T. Kunihiro, Phys. Lett. A **380**, 2075 (2016)
8. S.J. Barnett, Phys. Rev. **6**, 239 (1915)
9. S.J. Barnett, Rev. Mod. Phys. **7**, 129 (1935)
10. S.J. Barnett, 17 (n.d.)
11. M. Arabgol, T. Sleator, Phys. Rev. Lett. **122**, 177202 (2019)
12. M. Ono, H. Chudo, K. Harii, S. Okayasu, M. Matsuo, J. Ieda, R. Takahashi, S. Maekawa, E. Saitoh, Phys. Rev. B **92**, 174424 (2015)
13. D.T. Son, P. Surowka, Phys. Rev. Lett. 103, 191601 (2009)
14. D.E. Kharzeev, L.D. McLerran, H.J. Warringa, Nucl. Phys. A **803**, 227 (2008)

15. K. Fukushima, D.E. Kharzeev, H.J. Warringa, Phys. Rev. D 78, 074033 (2008)
16. Z.T. Liang, X.N. Wang, Phys. Rev. Lett. 94, 102301 (2005)
17. F. Becattini, F. Piccinini, J. Rizzo, Phys. Rev. C 77, 024906 (2008)
18. J. Dey, S. Satapathy, P. Murmu, S. Ghosh, arXiv:1907.11164 [hep-ph]

Chapter 19
Investigation of Elastic and Dynamical Properties of RhTiSb

Dipangkar Kalita and Atul Saxena

Abstract We have investigated the structural, elastic, and dynamical properties of RhTiSb alloy using the reliable and trending plane-wave basis sets within the density functional theory. The exchange correlation effects are treated by using the most suitable generalized gradient approximation (GGA) under Perdew–Burke–Ernzerhof (PBE) scheme. The electron–ion interactions have been treated with the Rappe–Rabe–Kaxiras–Joannopoulos ultrasoft potentials. The alloy is stable in α-phase with nonmagnetic (NM) state. The computed elastic constants (C_{ij}s) satisfy the mechanical stability criteria. The compound reveals the ductile nature with weak ionic behaviour and anisotropic nature. In the entire Brillouin region, the positive phonon frequencies are observed which suggests the dynamic stability of the studied alloy, and the phonon modes are found to be both Raman and infrared active.

19.1 Introduction

Half-Heusler alloys (HHAs) are ternary intermetallic compounds with the chemical formula of XYZ; the *d* transition metal elements are generally X and Y, and Z is the leading *sp* element group [1]. These compounds have a $C1_b$ type cubic structure with space group F4-3m. Nowadays, HHAs have received growing interest for different applications including spintronics [1, 2], photovoltaics, optoelectronics [3], topological insulators [4], and thermoelectric materials [5] due to their wide variety of enjoyable physical properties [6, 7]. It was de Groot et al. who investigated the properties of NiMnSb and PtMnSb for the very first time and found the alloys to exhibit half-metallic character i.e. one electron spin channel shows metallic behaviour at the

Electronic supplementary material The online version of this chapter (https://doi.org/10.1007/978-981-16-5407-7_19) contains supplementary material, which is available to authorized users.

D. Kalita (✉) · A. Saxena
Department of Physics, North-Eastern Hill University, Shillong 793022, India

R. G. Nair et al. (eds.), *Proceedings of 28th National Conference on Condensed Matter Physics*, Springer Proceedings in Physics 269,
https://doi.org/10.1007/978-981-16-5407-7_19

Fermi energy level and in contrast, the other electron spin channel shows the semiconductor or insulator character [8]. Subsequently, the half-metallic ferromagnetism of NiMnSb alloy was verified experimentally [9], and later, half–metallicity was also observed in many HHAs [9–12].

In HAs in general, it has been found that the total number of electrons in the valence shell and lattice dynamics play a significant role in determining their physical properties [13, 14]. Kaur et al. used generalized gradient approximation (GGA) to study the electrical, phononic, and thermoelectric properties of the HfRhSb alloy and discovered it to be a semiconductor [15]. A similar approach used by Coban et al. to study the structural, mechanical, electronic, optical, and vibrational properties of HfXSb (X = Co, Rh, Ru) alloys and found a metallic behaviour for HfRuSb [16].

So far in the case of RhTiSb HHA, only the structural properties have been investigated experimentally [17] and to the best of our knowledge, no literature on the study of elastic constants and dynamical properties of this compound are available. Therefore, a number of theoretical researchers were fascinated by this compound [18–22]. Ma et al. explored 378 HHAs including the RhTiSb alloy and reported it to be a nonmagnetic semiconductor in cubic structure with a lattice parameter of 6.12 Å and a bandgap of 0.75 eV [18]. Shi et al. also examined 27 new HHAs including RhTiSb using GGA and Heyd–Scuseria–Ernzerhof (HSE06) hybrid function to obtain a bandgap of 0.74 eV and 1.47 eV respectively [19]. Anissa et al. displayed the RhTiSb alloy using GGA and mBJ approximations and found it to be an excellent thermoelectric material [20]. Benzoudji et al. also investigated the effects of M being replaced by Ti, Zr, and Hf atoms in MRhSb compound using LDA (local density approximation) and GGA approximations, where all the compounds revealed ductile nature [21]. A similar GGA approximation had been done by Surucu et al. for predicting the structural, electronic, mechanical and lattice dynamical properties of XRhSb (X = Ti and Zr) compounds [22]. Apart from the above studies, very few studies have been done on the mechanical and lattice dynamical properties of RhTiSb. Therefore, in these contexts, we investigated the mechanical and dynamical properties of Rh-based half Heusler alloy. The compound's dynamic property is crucial as it plays an important role in determining the electron–phonon interaction and other vibrational properties. For future investigations of the compound, this present work may serve as a reference guide.

19.2 Computational Details

We have used the plane wave basis sets within the density functional theory (DFT) distributed with the Quantum ESPRESSO package for calculations in the present work [23]. The exchange–correlation effects were treated by using the most suitable generalized gradient approximation (GGA) under Perdew–Burke–Ernzerhof (PBE) scheme with Rappe–Rabe–Kaxiras–Joannopoulos ultrasoft potentials [24]. The energy convergence was achieved by expanding the basis set to a cutoff value of 75 Ry. For the integration in the first Brillouin zone, a $12 \times 12 \times 12$ automatically generated grid k-points was taken following the Monkhorst and Pack scheme. The convergence tolerance for the calculation of energy in successive iteration steps was set to 10^{-10} Ry. The mechanical and dynamical properties were investigated by the thermo_pw package interfaced with the Quantum-ESPRESSO Package [25]. For the calculation of mechanical properties of the studied sample, stress–strain method was implemented [26]. We used the q-point grid of 2 × 2 × 2 for calculating dynamical matrices.

19.3 Result and Discussion

19.3.1 Structural and Elastic Properties

Depending upon the number of possible combinations of atoms occupying the different Wyckoff positions in the unit cell, the crystal structure may be categorized into three phases: α-phase, β-phase, and γ-phase [13, 27]. The total energy as a function of lattice parameter in the ferrimagnetic (FiM), ferromagnetic (FM), and non-magnetic (NM) states for all the above 3 phases have been fitted to Murnaghan's equation of state as shown in Fig. 19.1a. Among the three phases, the α-phase with NM state is found to be energetically more favourable configuration where Rh atom sits at (0.25, 0.25, 0.25), Ti atom at (0.5, 0.5, 0.5), and Sb atom at (0, 0, 0) [21] as shown in Fig. 19.1b. The computed structural properties such as lattice parameter (a), bulk modulus (*B*), and its pressure derivative (*B'*) for this alloy are summarized in Table 1S (S = supplementary material). Our evaluated values are in good agreement with the other experimental [17] and analytical results [19–22].

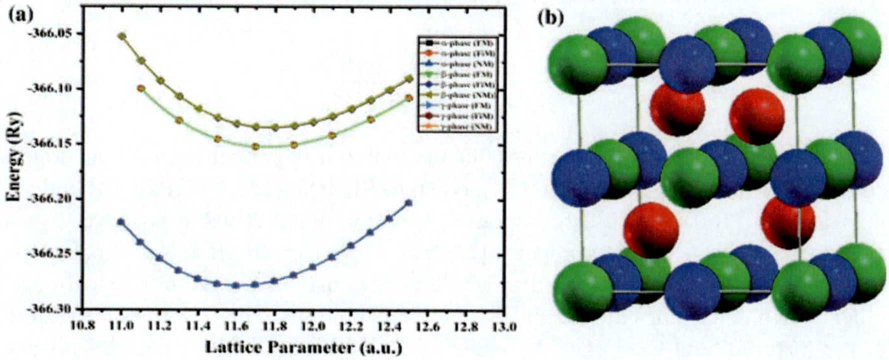

Fig. 19.1 **a** Volume optimization curve in the ferrimagnetic (FiM), ferromagnetic (FM) and nonmagnetic (NM) state for α-phase, β-phase, and γ-phase **b** Stable α-phase of the crystal structure (Rh = blue, Ti = red and Sb = green)

Table 19.1 Calculated independent elastic constant (C_{ij}), Young's modulus (Y), bulk modulus (B), Shear modulus (G) are in GPa, Anisotropic ratio (A), Poisson ratio (ν), Kleinmann parameter (ζ), Pugh ratio (*B/G*) and Debye temperature (θ_D) in K of the RhTiSb alloy in α-phase

RhTiSb compound	C_{11}	C_{12}	C_{44}	*Y*	*B*	*G*	*A*	ν	ζ	*B/G*	Θ_D
Present	209.23	108.29	61.64	150.63	141.94	56.92	1.22	0.32	0.64	2.49	335.66
Prev. [21]	208.66	110.79	94.11	187.73	143.42	73.23	1.92	0.28	–	1.95	–
Prev. [22]	208.77	108.85	57.68	144.87	142.16	54.46	1.16	0.33	–	2.61	328.37

To examine the structural stability and mechanical properties of the material, elastic constants are also computed. Since the studied sample crystallizes in face-centered cubic (FCC) lattice, there are only three independent elastic constants namely C_{11}, C_{12} and C_{44} as listed in Table 19.1. These C_{ij}s are found to satisfy the born stability criteria [28] and other physical parameters can be calculated as listed in Table 19.1.

The average value of Bulk modulus (B) and the Shear modulus (G) were estimated using the Voigt–Reuss–Hill approximation as [29, 30]

$$B = \frac{(C_{11} + 2C_{12})}{3} \tag{19.1}$$

$$G = \frac{(G_V + G_R)}{2} \tag{19.2}$$

where,

$$G_V = \frac{1}{5}(C_{11} - C_{12} + 3C_{44})$$

$$G_R = \frac{5(C_{11} - C_{12})C_{44}}{3(C_{11} - C_{12}) + 4C_{44}}$$

The B which defines the hardness of the material is greater than the analogous compound RhZrSb (127.92 GPa) [31], equal to RhHfSb (143.62 GPa) [15] and less than RhVSb (160 GPa) [32] compound. Our computed B value is in very good agreement with the previous reports [21, 22]. Moreover, the B value as calculated from the volume optimization curve and elastic constants are consistent with each other, which determines the reliability of our calculations. The G defines the rigidity of the material, and the larger the value of G, the higher is the rigidity. This also means that a large force is required to cause a deformation. From Table 19.1 we can see that $B > G$, indicating the prominence of shear modulus in the sample's stability.

From the above B and G values, one can determine Young's modulus (Y) and Poisson ratio (ν), which are described as follows [33]:

$$Y = \frac{9BG}{3B + G} \tag{19.3}$$

$$\nu = \frac{3B - 2G}{2(3B + G)} \tag{19.4}$$

Y is the stress to strain ratio and measures the stiffness of the material. The higher the value of Y, the greater will be the stiffness of the material. Our simulated Y result is slightly greater than that reported by Surucu et al. and less than the value obtained by Benzoudji et al. These differences are mainly–because of the dependence of elastic properties on the independent elastic constants. ν is a parameter that defines the form of the bonding of the substance, which indicates covalent bonding for $\nu = 0.1$ and ionic bonding for $\nu = 0.25$ [34]. In that case, the RhTiSb alloy with ν of 0.32 shows a weak ionic nature. The brittle and ductile properties of materials can also be distinguished by Pugh's ratio (*B/G*) [35]. If the value is greater than 1.75, the material is ductile, otherwise brittle. Table 19.1 reveals the ductile nature of RhTiSb alloy, which is comparable with other analogous compounds like RhZrSb (Ductile) [31], RhVSb (Brittle) [32], RhHfSb (Ductile) [15, 21]. Both ν and *B/G* are in good agreement with the previous reports [21, 22].

The anisotropy factor (A) and Kleinman parameter (ζ) have also been calculated as follows:

$$A = \frac{2C_{44}}{C_{11} - C_{12}} \tag{19.5}$$

$$\zeta = \frac{C_{11} + 8C_{12}}{7C_{11} - 2C_{12}} \tag{19.6}$$

A is an essential property of the material that reveals the compounds' structural stability and the deviation from unity indicates the possibility of developing micro-cracks or faults during the crystal growth process [32, 36]. If the value of $A = 1$, then the alloy is an isotropic system and conversely, deviation from 1, as in the present case indicates the anisotropic nature [37]. The next mechanical parameter is the Kleinman parameter (ζ), which explains the relative positions of the cation and anion sub-lattices under the strain distortion to conserve the volume which is not fixing the symmetry of the crystal structure. This parameter also defines the compound's structural stability [1]. Its value lies between 0 and 1 [38]. The calculated value of $\xi \approx 0.7$ means that the atomic positions are slightly stiff against the lattice's distortions.

One of the most important parameter is Debye temperature (θ_D) which is associated with the material's thermal properties. Using the elastic constants data, θ_D can be computed as:

$$\theta_D = \frac{h}{K_B}[\frac{3n}{4\pi} \times \frac{N_A\rho}{M}]^{\frac{1}{3}} \times V_m \tag{19.7}$$

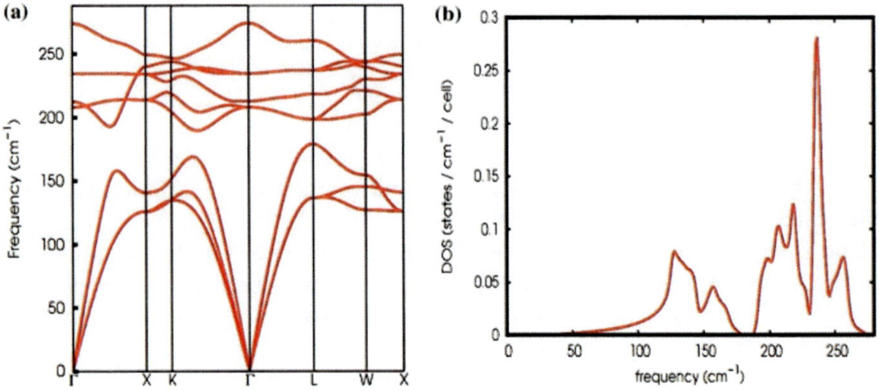

Fig. 19.2 (**a**) Phonon dispersion curve and (**b**) Phonon Density of States of RhTiSb alloy

Where h is Planck's constant, K_B is Boltzmann's constant and N_A, M, V_m and n are the Avogadro's number, molecular weight, average sound velocity, and the number of atoms per unit cell. The computed θ_D value indicates high lattice thermal conductivity and melting point compared to analogous compounds like RhZrSb (333.30 K) [31], HfRhSb (198.67 K) [16]. Our result is in reasonable agreement with the previous results [22].

19.3.2 Lattice Dynamical Properties

The stability of the phase of an alloy is determined by the lattice dynamical properties. The phonon dispersion curve along the high symmetric point $\Gamma - X - K - \Gamma - L - W - K$ of the Brillouin Zone (BZ) is shown in Fig. 19.2a. The absence of any imaginary phonon frequencies at the ambient condition means the stability of RhTiSb alloy in the cubic structure (α-phase). Since the sample alloy has three atoms in the primitive FCC unit cell, therefore there are nine phonon modes. Out of which, three phonon modes are acoustic and the remaining are optical modes. The irreducible representation for these optical phonon modes is represented as

$$\Gamma_{optical} = 2T_2 \tag{19.8}$$

Here, T_2 is found to be both infrared and Raman active and the T mode means that it has triple degeneracy.

In Fig. 19.2a, the acoustic and optical modes are separated from each other. Therefore, there is a phonon bandgap of 11.1 cm^{-1}. Though it is noticeable that there are two acoustic and four optical phonon modes present along 'Γ-L' (right) high-symmetry direction, and the three acoustic and six optical phonon modes are visible in the 'Γ-L' (left) direction.

19.4 Conclusion

The structural, elastic, and dynamical properties of RhTiSb were investigated using the plane-wave method. The sample alloy was found to be stable in the non-magnetic α phase. The mechanical and cubic stability condition was confirmed from the computed values of elastic constants with its ductile nature. The computed value of bulk modulus, 141.94 GPa was found to be in good agreement with previous results. The observed positive phonon modes confirmed the dynamical stability of RhTiSb with a phonon band gap of 11.1 cm^{-1}. The computed value of θ_D (335.66 K) indicates that the material posses strong bonding between the constituent atoms as compared to other analogous materials, RhZrSb and HfRhSb. The present work may pave the way to examine the pressure effect on the mechanical, thermodynamical and other related properties of RhTiSb.

References

1. M. Shakil, M. Kousar, S.S.A. Gillani, M. Rizwan, H. Arshad, M. Rafique, M. Zafar, Indian J. Phys. 1–12 (2020)
2. K. Elphick, W. Frost, M. Samiepour, T. Kubota, K. Takanashi, S. Hiroaki, A. Hirohata, Sci. Technol. Adv. Mater. (2020)
3. M. Naseri, D.M. Hoat, J. Mol. Graph. Model **92**, 249–255 (2019)
4. C.K. Barman, A. Alam, Phys. Rev. B **97**(7), 075302 (2018)
5. H. Joshi, D.P. Rai, A. Laref, R.K. Thapa, Mater. Res. Express **6**, 066307 (2019)
6. A. Erkisi, G. Surucu, R. Ellialtioglu, Phil. Mag. **97**(26), 2237–2254 (2017)
7. T. Graf, C. Felser, S.S.P. Parkin, Prog. Solid State Chem. **39**, 1–50 (2011)
8. R.A. De Groot, F.M. Mueller, P.G. Van Engen, K.H.J. Buschow, Phys. Rev. Lett. **50**, 2024–2027 (1983)
9. Z. Hao, R. Liu, Y. Fan, L. Wang, J. Alloys Compd. **820**, 153118 (2020)
10. A. Abada, N. Marbouh, J. Supercond. Novel Magn. **33**(3), 889–899 (2020)
11. M. Rostami, M. Abedi, P. Amantorkaman, F. Kanjouri, Vacuum **175**, 109278 (2020)
12. M. Baral, A. Chakrabarti, Phys. Rev. B **99**(20), 205136 (2019)
13. Y. Gupta, M.M. Sinha, S.S. Verma, Philos. Mag. 1–17 (2020)
14. D.M. Hoat, Comput. Mater. Sci. **159**, 470–477 (2019)
15. K. Kaur, R. Kumar, D.P. Rai, J. Alloys Compd. **763**, 1018–1023 (2018)
16. C. Çoban, Y.Ö. Çiftçi, K. Çolakoğlu, Indian J. Phys. **90**(11), 1233–1241 (2016)
17. A.E. Dwight, J. Less Common Metals **34**, 279–284 (1974)
18. J. Ma, V.I. Hegde, K. Munira, Y. Xie, S. Keshavarz, D.T. Mildebrath, W.H. Butler, Phys. Rev. B **95**(2), 024411 (2017)
19. F. Shi, M.S. Si, J. Xie, K. Mi, C. Xiao, Q. Luo, J. Appl. Phys. **122**, 215701 (2017)
20. B. Anissa, D. Radouan, B. Benaouda, Int. J. Mod. Phys. B **33**, 1950247 (2019)
21. F. Benzoudji, O.M. Abid, T. Seddik, A. Yakoubi, R. Khenata, H. Meradji, H.Y. Ocak, Chin. J. Phys. **59**, 434–448 (2019)
22. G. Surucu, A. Candan, A. Erkisi, A. Gencer, H. H. Güllü, Mater. Res. Express **6**(10), 106315 (2019)
23. P. Giannouzzi et al., J. Phys. Condens. Matter 21, 395502 (2009). https://www.Quantum-espresso.org
24. J. Perdew, K. Burke, M. Ernzerhof, Perdew, burke, and ernzerhof reply. Phys. Rev. Lett. **80**(4), 891 (1998)

25. A. Dal Corso, The Thermo_pw code is available at https://dalcorso.github.io/thermo_pw/
26. S.C. Wu, S.S. Naghavi, G.H. Fecher, C. Felser, J. Mod. Phys. **9**, 83535 (2018)
27. W. Huang, X. Wang, X. Chen, W. Lu, L. Damewood, C.Y. Fong, Mater. Chem. Phys **148**, 32–38 (2014)
28. M. Ram, A. Saxena, N. Limbu, H. Joshi, A. Shankar, J. Appl. Phys. **128**, 053901 (2020)
29. N. Limbu, A. Saxena, R.K. Thapa, A. Shankar, J. Solid State Chem. **282**, 121081 (2020)
30. R. Hill, Proc. Phys. Soc. Sect. A **65**, 349 (1952)
31. S.A. Khandy, J.D. Chai, J. Alloy. Compd. **850**, 156615 (2021)
32. R. Ahmad, N. Mehmood, J. Supercond. Novel Magn. **31**(5), 1577–1586 (2018)
33. Y. Li, Y. Gao, B. Xiao, T. Min, Z. Fan, S. Ma, L. Xu, J. Alloys Compd. **502**, 28 (2010)
34. M. Ram, A. Saxena, A.E. Aly, A. Shankar, RSC Adv. **10**, 7661–7670 (2020)
35. Y. Zhou, J. M. Zhang, Y.H. Huang, X.M. Wei, J. Phys. Chem. Solids **138**, 109246 (2020)
36. R. Ahmad, N. Mehmood, J. Supercond. Novel Magn. **31**(1), 257–264 (2018)
37. M.A. Bennani, Z. Aziz, S. Terkhi, E.H. Elandaloussi, B. Bouadjemi, D. Chenine, S. Bentata, J. Supercond. Novel Magn. 1–15 (2020)
38. S.A. Sofi, D.C. Gupta, J. Solid State Chem. 284, 121178 (2020)

Chapter 20
First Principles Study of TiO_2 as Visible Light Catalyst with Ni Doping

A. Angeline Dorothy and Puspamitra Panigrahi

Abstract Doping TiO_2 with noble metals, transition metals, cations, anions have yielded very promising results in enhancing photocatalytic activity of TiO_2 in the visible region and its role in generating alternate forms of energy. However, the study of Ni as a dopant can lead to a reliable and versatile TiO_2 modified photocatalyst, as Ni is an earth-abundant metal. In this paper we explore the electronic properties of Ni doped TiO_2 by varying the Ni dopant concentrations. As Ni doping increase from 4.17% to 8.33% to 12.5%, we observe Ni 3d orbitals being introduced into TiO_2 bandgap in the forbidden region and the hybridized orbitals of Ti 3d and Ni 3d extending well over the conduction band region near Fermi level indicating an increased metallic nature and hence conduction. The charge density plots confirm a metallic bond being developed as Ni concentration increases. The optical properties prove that Ni doped TiO_2 can absorb well in the visible region with an absorption coefficient of $1 \times 10^5\ \text{cm}^{-1}$. Hence Ni doped TiO_2 can successfully alter the electronic and optical properties of TiO_2 for favourable future applications.

20.1 Introduction

Doping TiO_2 with a variety of cations, anions, metals, and transition-metals, have been implemented in order to improve its efficiency for absorbing in the visible range by either decreasing the bandgap or by introducing intra-bandgap states [1, 2]. Further, the photocatalytic performance of TiO_2 is reported to increase by inhibiting the recombination of holes and electrons, when doped with noble metals like Au, Ag, and Pt. Among TiO_2 polymorphs, anatase with a wide bandgap of 3.2 eV has been proven to have a superior photocatalytic activity due to its longer electron–hole pair lifetime [3].

Other approaches like doping with transition metals too have been adopted to modify TiO_2 to absorb in the visible range. For instance Wang et al. have reported

A. A. Dorothy (✉) · P. Panigrahi
Hindustan Institute of Technology and Science, Chennai 603103, India
e-mail: rs.ad0914@hindustanuniv.ac.in

R. G. Nair et al. (eds.), *Proceedings of 28th National Conference on Condensed Matter Physics*, Springer Proceedings in Physics 269,
https://doi.org/10.1007/978-981-16-5407-7_20

the doping in TiO_2 with various transition metals [4]. They have reported that doping with Cr, Co, and Ni are energetically more favored at Ti site than O site. They have studied that the transition metal doping improves the photocatalytic activity of TiO_2 by narrowing down the bandgap to visible region. Doping with lanthanides have been reported to effectively enhance photocatalytic activity in visible region [5]. A few metals as such have also been studied as dopants and proved to be effective for band gap narrowing [6]. In general, noble metals are known to retard the carrier recombination and enhance photocatalytic activity. Recently, Alotaibi et al. [7] have reported that multifunctional thin films of TiO_2 doped with Cu show enhanced photocatalytic and antibacterial activity. In a similar work on Cr doped TiO_2 photoanodes, Song et al. attributed Cr 3d orbitals and Oxygen vacancies which produced a synergistic effect to increase visible light activity of Cr doped TiO_2 [8].

Jensen et al. [9] have studied Ni doped TiO_2 with H_2O molecules and deduced that Ni indeed reduces the bandgap of TiO_2 and enhances visible light photoactivity. Doping with Ni changes the phase composition and optical absorption range of TiO_2 film gradually expands and shifts to the red with increasing dosages enhancing the light utilization according to Yao et al. [10].

Being an earth abundant metal, we are well served in exploring Ni as a reliable and economical dopant for TiO_2. In this study we too have studied the doping of Ni into TiO_2 and using first principles studies investigated the density of states to ascertain the change in electronic structure caused by Ni. The absorption too has been noted to be enhanced by Ni doping in the visible region with a higher absorption coefficient in the visible region.

20.2 Computational Method

For this project, the calculations were ascertained by using first principle density functional theory (DFT) within Vienna Ab-initio Simulation Program [11] (VASP) package. We have used the Perdew–Burke–Ernzerhof (PAW-PBE) exchange functional [12] of the Generalized Gradient Approximation (GGA) [13] to perform the ground state structural optimization and electronic properties calculation. The energy cut-off for the plane wave basis set was optimised to be 550 eV. For ground state structure optimization, the Brillouin zone of the supercell was sampled by $3 \times 3 \times 3$ Monkhorst–Pack k-point mesh whereas for electronic structure calculation a higher mesh of $5 \times 5 \times 5$ was used. The DFT calculations are performed on a $2 \times 1 \times 1$ supercell of TiO_2 having 8 platinum and 16 oxygen atoms. As shown in Fig. 20.1, the substitutional doping concentration of Ni in Ti_8O_{16} is varied as 4.17%, 8.33% and 12.5% by substituting Ti atoms by 1, 2 and 3 Ni atoms respectively. The projected augmented wave (PAW) method was implemented, to explicitly treat their valence electrons. Both pristine Ti_8O_{16} and substitutional doped $Ti_{8-x}O_{16}Ni_x$, x = (1, 2, 3) structures were fully relaxed including lattice parameters and ionic positions until energies and forces reached convergence values. The criterion of electronic convergence in the self-consistent field was 10^{-7} eV and the force convergence was set to

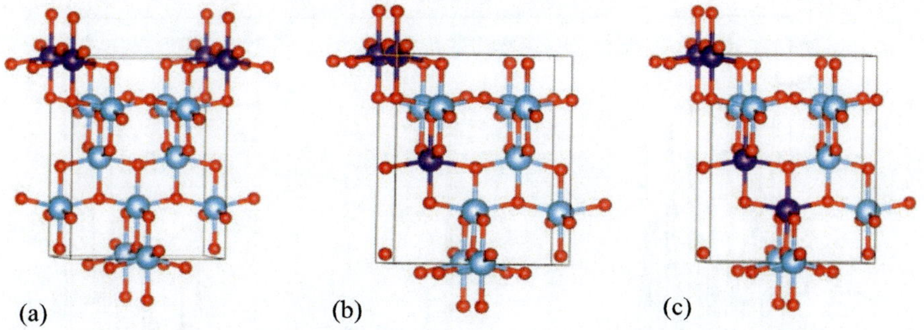

Fig. 20.1 The optimized structures of Ti_8O_{16} supercell doped with a 4.17% Ni b 8.33% Ni c 12.5% Ni. Colour: Blue: Ti, Red: Oxygen, Violet: Ni

be 0.001 eV/Å. For the optical properties, the frequency dependent dielectric matrix was calculated using VASP 5.4 optical programs. All the optical calculations were performed using $6 \times 6 \times 6$ grid mesh of Monkhorst–Pack scheme for the pure and 4.17% Ni doped TiO_2 structures.

20.3 Results and Discussion

In this section we discuss the various electronic and optical studies of Ni doped TiO_2. Figure 20.1 shows the optimized tetragonal structures of TiO_2 doped with 1, 2 and 3 Ni atoms respectively. In the following discussion we report the electronic structures of these doped structures.

20.3.1 Electronic Structure

The electronic density of states for TiO_2 doped with 1 Ni atom as in Fig. 20.2 shows a distinct valence band consisting of predominantly O 2p orbitals and the conduction band dominated by Ti 3d orbitals. The Fermi level is closer to the valence band and there is a distinct band gap seen. The conduction band shows presence of Ni 3d orbitals in the 1 eV region hybridizing well with the O 2p and Ti 3d orbitals. The peaks are distinct and the orbitals are far apart. The DOS affirms retention of semiconductor character in the structure.

However, the 2 Ni doped TiO_2 electronic structure shown in Fig. 20.3 reveals the Fermi level is now closer to the conduction band. The conduction band is seen to be overlapping with the valence band. The bandgap is closed and the structure has assumed a metallic structure. The forbidden region is now occupied by the Ni 3d orbitals distinctly. Strong hybridization of Ti 3d, O 2p and Ni 3d orbitals are seen in the conduction region of 0–2 eV.

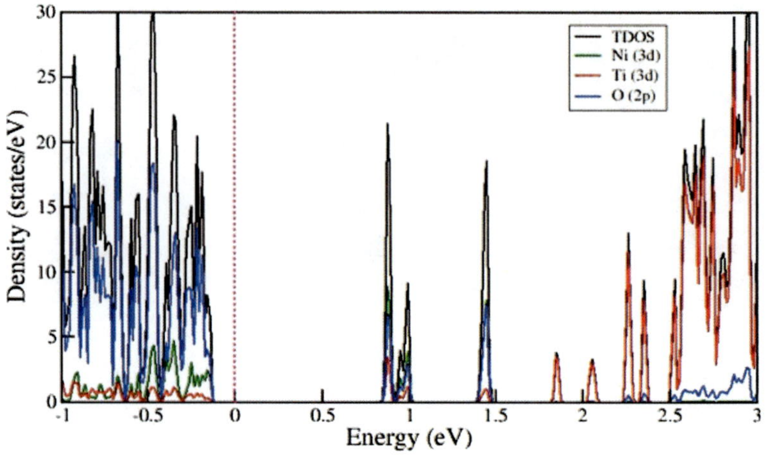

Fig. 20.2 Partial and total density of states of 1 Ni doped TiO_2.Pink line indicates Fermi level

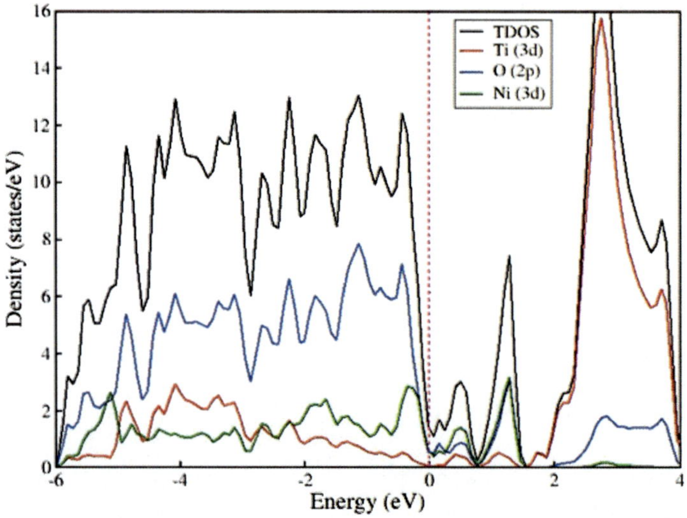

Fig. 20.3 Partial and total density of states of 2 Ni doped TiO_2. Pink line indicates Fermi level

Figure 20.4 for the 3 Ni atom doped structure reveals a more prominent presence of new states in the forbidden region with higher intensity of Ni 3d orbitals compared to the 2 Ni case, thus indicating a stronger metallic character with increased Ni doping. A closer look at the Ni 3d states show intensity of Ni 3d orbitals closer to the Fermi level has increased compared to the 2 Ni case. The forbidden region holds a higher intensity of exclusively Ni 3d orbitals. Similarly the CB region between 0–2 eV too shows that intensity of Ni 3d orbitals strongly exceeding over the Ti 3d orbitals.

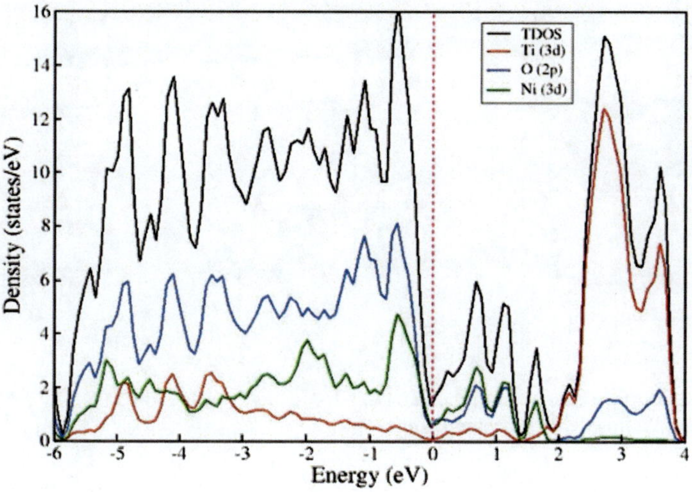

Fig. 20.4 Partial and total density of states of 3 Ni doped TiO_2. Pink line indicates Fermi level

Hence we can conclude that Ni doping has introduced new states in the forbidden region and the conduction band.

Thus we can expect a stronger presence of Ni 3d orbitals dominating the conduction band as Ni concentration increases giving it a metallic character and thus its conducting nature too.A similar trend has been reported by Jianning Li et al for Cr/S co-doped rutile TiO_2 where the forbidden region is occupied by Cr 3d and S 3p orbitals [14].

20.3.2 Charge Density

Figure 20.5 shows the charge density plots for the Ni doped TiO_2 structures.

The charge density plots show a charge depletion occurring both for the 1Ni atom and the corresponding O atom bonded to Ni and a covalent charge cloud over the Ni–O bond. However the remaining O atoms show charge accumulation as is normal. In the 2 Ni atom case the charge depletion of Ni is increased but the charges are not correspondingly depleted for the O bonded to Ni thus suggesting tendency toward a metallic bond. But the 3 Ni atom structure shows a stronger charge depletion occurring in the Ni atoms but significantly weaker charge depletion in the O bonded to them. So we can confirm that Ni is not only receiving a cationic character but the Ni–O bond is more metallic for the 2 Ni and 3 Ni cases than the covalent charge cloud in the 1 Ni case. That Ni doping causes the Ni site in TiO_2 to be a catalytic site is affirmed by Jensen et al. [9].

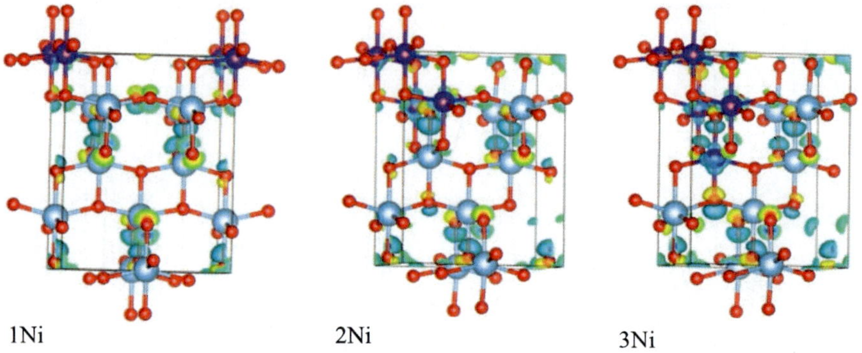

Fig. 20.5 Isosurface charge density of $Ti_5O_{16}Ni_{1-3}$ using isovalue of 0.005 e $Å^{-3}$. Cyan: charge depletion and Yellow: charge accumulation

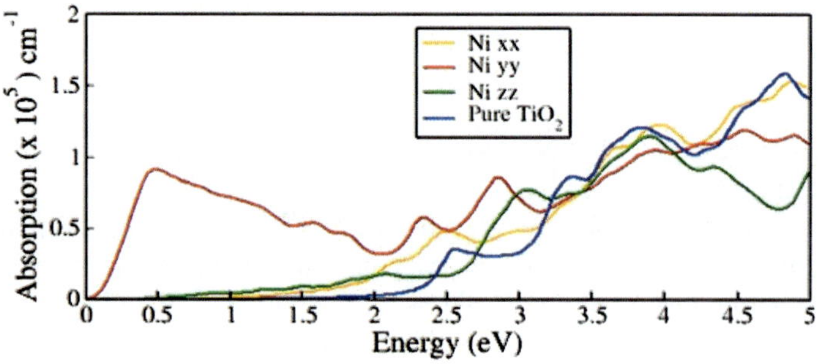

Fig. 20.6 The absorbance of Ni doped TiO_2

20.3.3 Optical Properties

Figure 20.6 shows the absorption spectra for Ni doped (4.17%) TiO_2 in all three directions compared with that of pure TiO_2. As can be seen, there is shift of the absorption towards the visible light region for the doped system. Though the absorption coefficient for pure system is higher in UV region compared to the visible region, the Ni doped TiO_2 shows high absorption of 1×10^5 cm^{-1} in the visible region also. Hence we can state that visible light absorption is enhanced with Ni doping in TiO_2 although the intensity is lower than that of pure TiO_2 in UV region.

20.4 Conclusions

The detailed study of electronic structure of Ni doped TiO_2 reveal the dramatic change from semiconducting nature to metallic nature as doping concentration increases from 4.17% to 8.33% to 12.5%. From 8.33% to 12.5% to Ni dopant levels, interestingly Ni 3d states are introduced closer to the Fermi level within the forbidden region. So at these concentrations, metallic character sets in. The charge density plots reveal a similar trend as concentrations vary. In the case of 1 Ni atom dopant, the charge density difference plots show a predictable cationic Ni atom with a charge depleted O atom But in the cases of 2 Ni and 3 Ni atoms, the Ni atoms show dramatic increase in charge depletion whereas a comparative lowering of charge depletion of the O atoms bonded to Ni atoms, thus authenticating a tendency that a metallic bond is being developed. The optical properties affirm that Ni doping enhances absorption in visible region although with lower intensity compared to that of pure TiO_2 in UV region.

Hence we conclude the enhancement in visible light photocatalytic activity and conduction ability of Ni doped TiO_2 which can have many photocatalytic applications.

References

1. A. Ghicov, B. Schmidt, J. Kunze, P. Schmuki, Photoresponse in the visible range from Cr doped TiO_2nanotubes. Chem. Phys. Lett. **433**, 323–326 (2007). https://doi.org/10.1016/j.cplett.2006.11.065
2. Z. Jiang, Y. Lin, T. Mei, X. Hu, W. Chen, R. Ji, E. Liu, R. Zhang, L. Zhang, Q. Zhang, B. Zhou, D. Zhang, J. Fan, H. Zhu, X. Zhang, S. Wan, S. Zhu, Y. Shang, First-principles study of the electronic and optical properties of the (Eu, N)-codoped anatase TiO_2 photocatalyst. Comput. Mater. Sci. **68**, 234–237 (2013). https://doi.org/10.1016/j.commatsci.2012.09.021
3. S. Rehman, R. Ullah, A.M. Butt, N.D. Gohar, Strategies of making TiO_2 and ZnO visible light active. J. Hazard. Mater. **170**, 560–569 (2009). https://doi.org/10.1016/j.jhazmat.2009.05.064
4. Y. Wang, R. Zhang, J. Li, L. Li, S. Lin, First-principles study on transition metal-doped anatase TiO_2.Nanoscale Res. Lett. **9,** 46 (2014).
5. W. Chen, P. Yuan, S. Zhang, Q. Sun, E. Liang, Y. Jia, Electronic properties of anatase TiO_2 doped by lanthanides: A DFTU study. Phys. B Condens. Matter. **407**, 1038–1043 (2012). https://doi.org/10.1016/j.physb.2012.01.085
6. R. Long, Y. Dai, G. Meng, B. Huang, Energetic and electronic properties of X- (Si, Ge, Sn, Pb) doped TiO_2 from first-principles. Phys. Chem. Chem. Phys. **11**, 8165–8172 (2009). https://doi.org/10.1039/b903298c
7. A.M. Alotaibi, B.A.D. Williamson, S. Sathasivam, A. Kafizas, M. Alqahtani, C. Sotelo-Vazquez, J. Buckeridge, J. Wu, S.P. Nair, D.O. Scanlon, I.P. Parkin, Enhanced photocatalytic and antibacterial ability of Cu-doped anatase TiO_2 thin films: theory and experiment. ACS Appl. Mater. Interfaces. **12**, 15348–15361 (2020). https://doi.org/10.1021/acsami.9b22056
8. X. Song, W. Li, X. Liu, Y. Wu, D. He, Z. Ke, L. Cheng, C. Jiang, G. Wang, X. Xiao, Y. Li, Oxygen vacancies enable the visible light photoactivity of chromium-implanted TiO_2 nanowires. J. Energy Chem. **55**, 154–161 (2021). https://doi.org/10.1016/j.jechem.2020.07.013
9. S. Jensen and D.S. Kilin, Electronic properties of nickel-doped TiO_2 anatase, J. Phys. Condens. Matter. **27,** 134207 (2015). https://doi.org/10.1088/0953-8984/27/13/134207

10. Z. Yao, F. Jia, S. Tian, C. Li, Z. Jiang, X. Bai, Microporous Ni-Doped TiO_2 film photocatalyst by plasma electrolytic oxidation. ACS Appl. Mater. Interfaces. **2**, 2617–2622 (2010). https://doi.org/10.1021/am100450h
11. G. Kresse and J. Furthmüller, Efficient iterative schemes for ab-initio total energy calculations using a plane-wave basis set. Phys. Rev. B.-Condens. Matter Mater.Phys. **54**, 11169-86 (1996)
12. J. P. Perdew,K. Burke and M.Ernzerhof, Generalized Gradient Approximation Made Simple, Phys. Rev. Lett. **77**, 3865–3868 (1996)
13. P.E.Blöchl, Projector augmented wave-method, Phys.Rev B. **50**,17953-79(1994)
14. J. Li, F. Wu, J. Shi, L. Ma, X. Yan, N. Yang, B. Ding, S. Zheng, First-principles calculation of Cr/S Co-doped rutile TiO_2. Mater. Sci. Pol. **38**, 253–262 (2020). https://doi.org/10.2478/msp-2020-0042

Chapter 21
Effect of Parametric Variation on Performance of NFA Organic Solar Cell: A Simulation Study

Shivam Dave, Santosh V. Patil, and Kshitij Bhargava

Abstract The organic bulk heterojunction non-fullerene acceptor (NFA) solar cells have attracted great interest among researchers as they provide a higher degree of stability and better efficiency over their fullerene-based counterparts. However, a shallow simulation effort has been made for understanding the impact of intrinsic parameters on performance of NFA based organic solar cells. This work provides a comprehensive study on a traditional p-i-n structure of a cell that utilizes (ITIC):3,9-bis (2-methylene(3- (1,1-dicyanomethylene) -indanone)—5,5,11,11-tetraki (4-hexyl phenyl)-di-thieno [2,3d:2,3-d]-s-indaceno [1,2-b:5,6 b'] di-thiophene) as NFA and poly [(2,6-(4,8-bis (5(2-Ethylhexyl) thiophene-2-yl) benzo [1,2-b:4,5b] di-thiophene)-co-(1,3-di (5-thiophene2-yl)-5,7-bis (2-Ethylhexyl) benz 1,2-c:4,5-c] di-thiophene-4,8-dione)] (PBDB-T) as polymeric donor. We have studied the aftermath of variation of various parameters of an organic solar cell on the power conversion efficiency, open-circuit voltage, short circuit current density, and fill factor of solar cell. The work is divided into three phases; in the first phase, the effect of the thickness of absorber layer on the performance parameters is analyzed. The second phase emphasizes on the effect of change in electron and hole mobility on solar cell efficiency and the third phase concentrates on the variation of bulk and interfacial defect density on performance parameters of the cell.

21.1 Introduction

Solar photovoltaic technology is rapidly growing however the limited power conversion efficiency of solar cells has been the bottleneck for its commercialization. The third-generation organic solar cells (OSCs) have become quite popular owing to their lucrative features like low-cost, lightweight, biodegradability, and flexibility [1, 2]. The device structure of an OSC consists of mainly an absorber layer, electron (ETL) and hole transport layers (HTL), front and back contacts. Generally, fullerene-based

S. Dave (✉) · S. V. Patil · K. Bhargava
Department of Electrical and Computer Science Engineering, Institute of Infrastructure, Technology, Research and Management, Ahmedabad, Gujarat 380026, India

R. G. Nair et al. (eds.), *Proceedings of 28th National Conference on Condensed Matter Physics*, Springer Proceedings in Physics 269,
https://doi.org/10.1007/978-981-16-5407-7_21

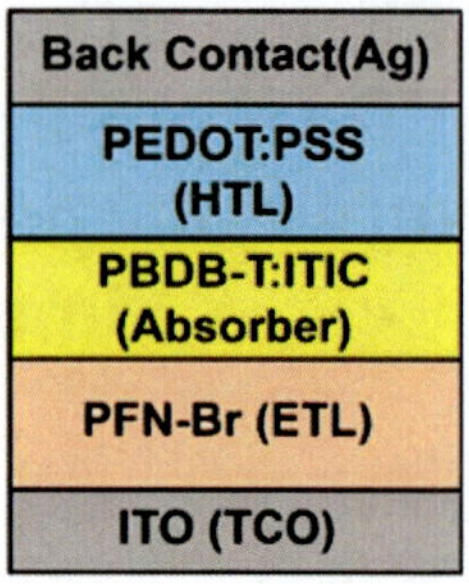

Fig. 21.1 Schematic representation of adopted model in the study

absorber materials are used to fabricate the OSCs but they fail to absorb the infra-red region of the solar spectra, have poor stability, and are expensive [3, 4]. In this context, non-fullerene acceptor (NFA) based absorber material, PBDB-T:ITIC has emerged as a more stable and inexpensive alternative since it absorbs both visible and the infra-red regions of the spectra [3, 5–8]. This study analyzes the impact of variation of parameters over the performance of PBDB-T:ITIC OSC through 1D simulations. The parametric variation is done in terms of absorber layer thickness, carrier mobility and defect density to achieve optimized performance of NFA OSC.

21.2 Simulation Methodology

The adopted model for the study is shown in Fig. 21.1 which represents a conventional p-i-n cell structure based on PBDB-T:ITIC absorber layer. PFN-Br was used as the ETL, PEDOT:PSS was chosen for HTL, Ag as back contact and ITO as front contact. The model was constructed using SCAPS-1D simulator which is a free software designed for simulating the thin film solar cells. The dimensions and parameters of the model were incorporated judiciously as summarized in Table 21.1 [5].

21.3 Results and Discussion

21.3.1 Effect of Absorber Layer Thickness

The thickness of absorber layer plays a crucial role in enhancing the performance of solar cell. Therefore, its thickness has to be optimized to achieve maximum efficiency without material wastage. For this purpose, the absorber layer's thickness (T) was varied from 50 ~ 200 nm and its impact on performance of cell was studied. The results shown in Figs. 21.2 and 21.3 demonstrate that the optimized thickness for absorber layer is 120 nm as the values of performance metrics of simulated cell become saturated at this point. At lower value of the thickness (50 ~ 100 nm),

Table 21.1 List of parameters for NFA OSC simulation

Parameters	PBDB-T:ITIC	PEDOT:PSS	PFN-Br	ITO
Thickness	120 nm	20 nm	5 nm	50 nm
Bandgap (eV)	1.2	1.6	2.98	3.3
Electron affinity (eV)	4.03	3.4	4	4.1
Dielectric permittivity	3.65	2.9	5	2
CB effective density of states (cm^{-3})	10^{19}	10^{22}	10^{19}	10^{20}
VB effective density of states (cm^{-3})	10^{19}	10^{22}	10^{19}	10^{19}
Electron and hole thermal velocity (cm/s)	10^{7}	10^{7}	10^{7}	10^{7}
Electron mobility (cm^2/V.s)	3×10^{-4}	4.5×10^{-4}	2×10^{-6}	5×10^{1}
Hole mobility (cm^2/V.s)	3×10^{-4}	9.9×10^{-3}	1×10^{-4}	7.5×10^{1}
Shallow uniform donor density (N_D) (cm^{-3})	0	0	9×10^{18}	10^{19}
Shallow uniform acceptor density (N_A) (cm^{-3})	10^{19}	2×10^{18}	0	0
Defect Density N_t (cm^{-3})	10^{12}	10^{9}	10^{9}	10^{12}

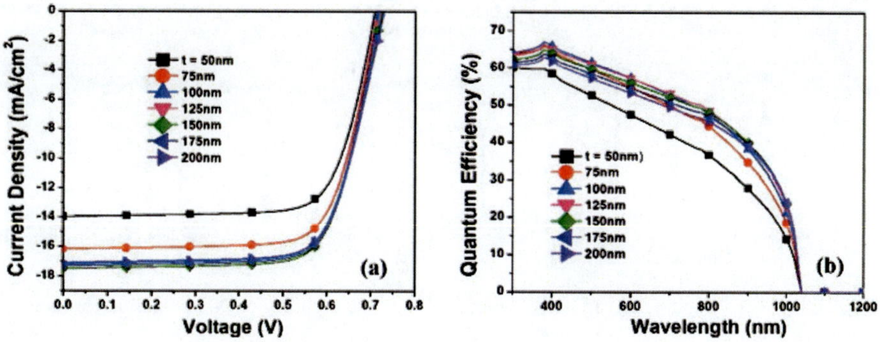

Fig. 21.2 Plot of **a** current density versus voltage **b** quantum efficiency versus wavelength characteristics for thickness optimization

the photons are not completely absorbed resulting in reduced photogeneration and thereby poor efficiency. A maximum of 9.15% efficiency was obtained at 120 nm thickness.

21.3.2 Effect of Carrier Mobility

The carrier mobility plays an important part in enhancing the performance of OSCs which in turn is governed by the crystallinity of absorber layer [9, 10]. Carrier

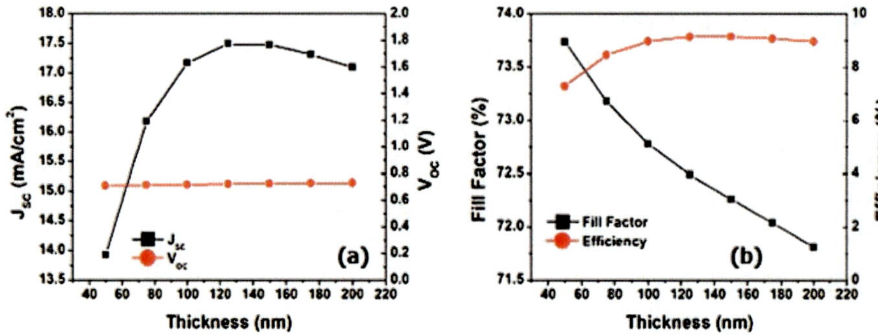

Fig. 21.3 Plot of **a** short-circuit current density/open-circuit voltage versus thickness **b** fill-factor/efficiency versus thickness for thickness optimization

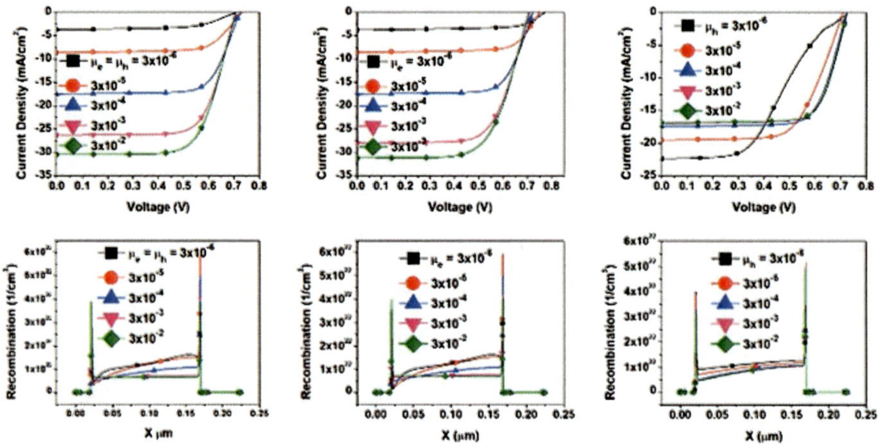

Fig. 21.4 Plots studying the variation of electron and hole mobility on NFA OSC performance

mobility plays a significant role in the separation and transportation of charges to the electrodes. In this subsection, we will study the importance of improving crystallinity of NFA OSC performance (Fig. 21.4).

21.3.2.1 Effect of Simultaneous Variation of Electron and Hole Mobility

The electron and hole mobility (μ_e and μ_h) were varied simultaneously (balanced condition) in the range 10^{-6} ~10^{-2} cm^2/Vs. It can be observed that with increasing mobility, there was significant rise in J_{SC} although V_{QC} remained consistent which is attributed to increased carrier collection at electrodes. These observations are well supported by the trends of fill-factor and efficiency. At lower mobility values, the recombination rate of the excitons increased causing increment in dark current and thus poor photocurrent generation.

21.3.2.2 Effect of Variation of Electron Mobility

In this part, the electron mobility $\mu_e$was varied from 10^{-6} to 10^{-2} cm^2/Vs, while hole mobility μ_h was maintained constant at a value of 10^{-4} cm^2/Vs. From the plots, it can be seen that the V_{OC} decreases with an increase in μ_e, while J_{SC} rises sharply.

21.3.2.3 Effect of Variation of Hole Mobility

Here, the hole mobility μ_h was varied from 10^{-6} to 10^{-2} cm^2/Vs, while the hole mobility μ_e was maintained constant at a value of 10^{-4} cm^2/Vs. And with an increase in the mobility of holes, V_{OC} rises while J_{SC} decreases, which means that with high values of hole mobility, the V_{OC} is aided.

21.3.3 Effect of Defect Density

21.3.3.1 Effect of Bulk Defect Density Variation

In this section, the defect density in the absorber layer was varied from 10^{10} to 10^{15} cm^{-3}. With high defect density, more traps are introduced within the layer thus causing enhanced recombination of the charge carriers and increases recombination. Due to high recombination, J_{SC} reduces and the cell's efficiency is reduced.

21.3.3.2 Effect of Absorber/ETL Interface Defect Density (IL_1) Variation

The defect density was varied from 10^7 to 10^{11} cm^{-3}. There was not much variation in J-V characteristics due to enhanced recombination rate at ETL/Absorber interface. The analysis reveal that it is important of reduce the defect density concentration inside bulk and at interfaces of absorber layer. The effect of defect density variation is depicted in Fig. 21.5.

21.3.3.3 Effect of HTL/Absorber Interface Defect Density (IL_2) Variation

Defects in the interfacial layers are formed due to the lattice mismatch of two layers, due to which reduced mobility of charge carriers is observed at the interface. Here

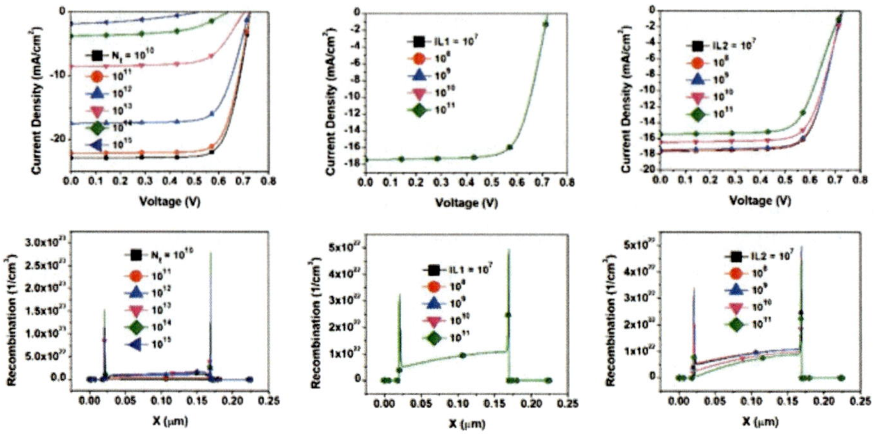

Fig. 21.5 Plots studying the variation of defect density on NFA OSC performance

the defect density was varied from 10^7 to 10^{11} cm^{-3}. Due to reduced mobility, the recombination rate rises, thus lowering the cell performance (Tables 21.2 and 21.3).

21.4 Conclusion

In this report, the effect of variation in intrinsic parameters on the performance of NFA solar cell was studied. The NFA based solar cell show a greater quantum efficiency (upto 1050 nm) as compared to that of the conventional P3HT:PCBM absorber that fail to absorb photons post 700 nm.The analysis revealed that the efficiency of simulated cell first increases with thickness and later decreases due to dominant effect of recombination. When electron and hole mobility were varied, it was observed that with the increased mobility, the efficiency increased due of the reduced extent of recombination. Moreover, rise in J_{SC} with electron mobility and V_{OC} with hole mobility were observed. Further, in this study, the increased defect density leads to rise in the rate of recombination. The performance of NFA based solar cell can be enhanced if the mobility of carriers improve. Simulation of NFA based absorber with different charge transport layers can also provide insights for better configuration of materials.

Table 21.2 Performance parameters for variation in mobility

Parameters	$N_t = 10^{10} - 10^{15}$; $IL_1 = IL_2 = 10^9$						$IL_1 = 10^7$ - 10^{11}; $IL_2 = 10^9$; $N_t = 10^{12}$					$IL_2 = 10^7$ - 10^{11}; $IL_1 = 10^9$; $N_t = 10^{12}$				
	10^{10}	10^{11}	10^{12}	10^{13}	10^{14}	10^{15}	10^7	10^8	10^9	10^{10}	10^{11}	10^7	10^8	10^9	10^{10}	10^{11}
V_{OC}	0.72	0.72	0.72	0.70	0.64	0.51	0.72	0.72	0.72	0.72	0.72	0.72	0.72	0.72	0.73	0.73
J_{SC}	22.9	22.2	17.48	8.60	3.77	1.85	17.48	17.48	17.48	17.48	17.48	17.66	17.64	17.48	16.49	15.47
FF	75.5	74.9	72.2	66.5	53.5	36.7	72.2	72.2	72.2	72.2	72.2	72.4	72.3	72.2	70.9	66.6
η	12.6	12.1	9.1	4	1.3	0.3	9.1	9.1	9.1	9.1	9.1	9.2	9.2	9.1	8.5	7.5

Table 21.3 Performance parameters for variation in defect density

Parameters	$\mu_e = \mu_h$					$\mu_e = 10^{-6} - 10^{-2}$; $\mu_h = 10^{-4}$					$\mu_h = 10^{-6} - 10^{-2}$; $\mu_e = 10^{-4}$				
	10^{-6}	10^{-5}	10^{-4}	10^{-3}	10^{-2}	10^{-6}	10^{-5}	10^{-4}	10^{-3}	10^{-2}	10^{-6}	10^{-5}	10^{-4}	10^{-3}	10^{-2}
V_{OC}	0.71	0.73	0.72	0.71	0.70	0.77	0.75	0.72	0.70	0.70	0.69	0.71	0.72	0.72	0.72
J_{SC}	3.75	8.66	17.48	26.34	30.42	3.81	8.62	17.48	27.98	31.17	22.35	19.57	17.48	16.95	16.91
FF	60.4	69.2	72.2	70.7	68.8	66.4	72.3	72.2	67.7	65.5	45.95	64.7	72.2	73.5	73.7
η	1.6	4.4	9.1	13.2	14.8	1.9	4.6	9.1	13.4	14.4	7.10	9.01	9.15	9.07	9.07

Acknowledgements Author K.B. is grateful to Dr. Marc Burgelman, Department of Electronics and Information Systems (ELIS), University of Gent, Belgium, for providing the SCAPS-1D simulation software.

References

1. O.A. Abdulrazzaq, V. Saini, S. Bourdo, E. Dervishi, A.S. Biris, Organic solar cells: a review of materials, limitations, and possibilities for improvement. Part. Sci. Technol. **31**(5), 427–442 (2013)
2. N Marinova, S Valero, J.L. Delgado, Organic and perovskite solar cells: Working principles, materials and interfaces. J. Colloid Interface Sci. **488**, 373–389 (2017)
3. K.S. Nithya, K.S. Sudheer, Device modelling of non-fullerene organic solar cell with inorganic CuI hole transport layer using SCAPS 1-D, Optik - Int. J. Light. Electron Opt. **217**, 164790 (2020)
4. W. Abdelaziz, A. Zekry, A. Shaker, M. Abouelatta, Numerical study of organic graded bulk heterojunction solar cell using SCAPS simulation. Sol. Energy **211**, 375–382 (2020)
5. W. Abdelaziz, A. Shaker, M. Abouelatta, A. Zekry, Possible efficiency boosting of non-fullerene acceptor solar cell using device simulation. Opt. Mater. **91**, 239–245 (2019)
6. C. Bendenia, H. Merad-Dib, S. Bendenia, Numerical modelisation of ZnO interfacial layer on P3HT:PCBM based organic photovoltaic bulk heterojunction devices. Optik **174**, 167–172 (2018)
7. Z. Zheng, Q. Hu, S. Zhang, D. Zhang, J. Wang, S. Xie, R. Wang, Y. Qin, W. Li, L. Hong, N. Liang, A highly efficient non-fullerene organic solar cell with a fill factor over 0.80 enabled by a fine-tuned hole-transporting layer. Adv. Mater. **30**(34), 1801801 (2018)
8. K.S. Nithya, K.S. Sudheer, Numerical modelling of non-fullerene organic solar cell with high dielectric constant ITIC-OE acceptor. J. Phys. Commun. **4**(2), 025012 (2020)
9. S. Khelifi, E. Voroshazi, D. Spoltore, F. Piersimoni, S. Bertho, T. Aernouts, J. Manca, J. Lauwaert, H. Vrielinck, M. Burgelman, Effect of light induced degradation on electrical transport and charge extraction in polythiophene: fullerene (P3HT: PCBM) solar cells. Sol. Energy Mater. Sol. Cells **120**, 244–252 (2014)
10. J.-T. Shieh, C.-H. Liu, H.-F. Meng, S.-R. Tseng, Y.-C. Chao, S.-F. Horng, The effect of carrier mobility in organic solar cells. J. Appl. Phys. **107**, 084503 (2010)

Chapter 22
Comparative Analysis of $MAPbI_3$ and $FAPbI_3$ based Perovskite Solar Cells: A Numerical Evaluation

Santosh V. Patil, Shivam Dave, and Kshitij Bhargava

Abstract In this paper, the influence of absorber defect density and acceptor doping concentration variation is analyzed in terms of J-V, QE-λ, bandgap alignment, and recombination rate to compare the performance of methylammonium lead halide ($MAPbI_3$) and formamidinium lead halide ($FAPbI_3$) solar cells. For this, two conventional planar type architectures viz. Glass/ZnO/$MAPbI_3$/Spiro-OMeTAD/Au and Glass/ZnO/$FAPbI_3$/Spiro-OMeTAD/Au were simulated and compared using SCAPS-1D. The analysis of simulation results revealed that the $FAPbI_3$ cell showed better performance with maximum power conversion efficiency of 21.26%, open-circuit voltage of 0.9989 V, short circuit current of 26.75 mA/cm^2, and fill factor of 79.80%. While, $MAPbI_3$ cells showed maximum power conversion efficiency of 18.17%, open-circuit voltage 0.9784 V, short circuit current of 23.75 mA/cm^2, and fill factor of 78.17%. This comparison provides information about the best alternative thin-film, perovskite solar cell (PSC) having high efficiency.

22.1 Introduction

The general structure of inorganic–organic lead halide perovskite is ABX_3, where A is the organic cation $CH_3NH_3^+$ (MA^+) or $NH_2CHNH_2^+$ (FA^+), B is for divalent metal cation (Pb^+ or Sn^+), and X is a halide anion (I^-) [1]. $MAPbI_3$ has an excellent optical property with a bandgap of 1.57 eV [2] that covers a wide range of visible light spectrum with a high absorption coefficient of 1.26×10^5 cm^{-1}. However, the limited light absorption in the IR region, structural imperfections, and thermal instability are some of the shortcomings of $MAPbI_3$ which leads to low crystallization energy and low-temperature phase adaption [3]. The aforementioned essential properties of $MAPbI_3$ are promoting researchers towards conceivable alternative absorbers. The

S. V. Patil (✉) · S. Dave · K. Bhargava
Department of Electrical and Computer Science Engineering, Institute of Infrastructure, Technology, Research and Management, Ahmedabad 380026, Gujarat, India

K. Bhargava
e-mail: kshitijbhargava@iitram.ac.in

R. G. Nair et al. (eds.), *Proceedings of 28th National Conference on Condensed Matter Physics*, Springer Proceedings in Physics 269,
https://doi.org/10.1007/978-981-16-5407-7_22

formamidinium lead iodide ($FAPbI_3$) has acquired greater attention as a perovskite absorber than traditional $MAPbI_3$ since it has an optimal bandgap (1.47 eV) [2, 4] closer to the infrared (IR) region and a high optical absorption coefficient of 1.62×10^5 cm^{-1}. Additionally, higher carrier diffusion length and more stability than $MAPbI_3$ have made $FAPbI_3$ a fascinating choice and appealing alternative of $MAPbI_3$ absorbers in PSCs [4]. The main advantage of $FAPbI_3$ over the $MAPbI_3$ is that it forms a more uniform and continuous thin film with fewer pinholes or defects [3]. In this study, the performance of $MAPbI_3$ and $FAPbI_3$ based PSCs with ZnO as electron transport layer (ETL) and Spiro-OMeTAD as hole transport layer (HTL) using SCAPS-1D. The analysis of J-V characteristic, Quantum Efficiency (QE vs. λ), Energy band diagram (E-x), and Recombination rate (RR vs. depth of the surface) are carried out to compare their performances. Furthermore, the effect of absorber defect density and doping concentration variations on $MAPbI_3$ and $FAPbI_3$ PSCs are also compared.

22.2 Modeling and Simulation Details

Two n-i-p type planar structures, FTO/ZnO/$MAPbI_3$/Spiro-OMeTAD/Au and FTO/ZnO/$FAPbI_3$/Spiro-OMeTAD/Au were modeled (Fig. 22.2). Table 22.1, shows the input simulation parameters incorporated in the SCAPS simulator. The SCAPS (Solar Cell Capacitance Simulator) is a one-dimensional simulation program with seven semiconductor input layers developed by a group of solar cell researchers at the Department of Electronics and Information System, University of Gent, Belgium. The SCAPS numerical modeling evaluates the performance of the solar cell by solving the equivalent mathematical Poisson's and continuity equations for holes and electrons iteratively. The back contact, gold (Au) is used with the work function equal to 5.1eV. Figure 22.1 shows the optical absorption model set in the SCAPS-1D software and expressed by the equation as,

$$\alpha(\lambda) = (A + \frac{B}{hv})\sqrt{hv - E_g} \tag{22.1}$$

where, E_g is the bandgap profile of the absorber material; A and B are the model parameters in cm^{-1} eV$^{-1/2}$, for the perovskite solar cell $A = 10^{+5}$, $B = 0$ [7].

Figure 22.1 shows the UV–vis absorption spectra of $MAPbI_3$ and $FAPbI_3$ perovskite films. In the range of wavelength 300–860 nm, the UV–vis absorption of $FAPbI_3$ perovskite film is stronger than that of $MAPbI_3$ in the range of 300 to 780 nm. Moreover, the different semiconductor materials and processing techniques can be developed to solve the stability problem. The decomposition temperatures used for $MAPbI_3$ and $FAPbI_3$ were 320° and 337° C, indicating that these films are still intact after annealing at 100 °C and 150 °C for 15 min respectively. Interestingly, the $FAPbI_3$ layer shows high sublimation temperature and quite higher charge diffusion length

Table 22.1 Simulation parameters incorporated in SCAPS

Parameters	Spiro-OMeTAD [3]	$MAPbI_3$ [5]	$FAPbI_3$ [3]	ZnO [5]	FTO [5]
Thickness (μm)	0.35	0.6	0.6	0.35	0.15
Bandgap Energy (Eg) eV	2.9	1.57 [2]	1.47 [2, 2]	3.47	3.5 [6]
Electron affinity (γ) eV	2.1	3.9	4	4.3	4 [6]
Relative dielectric permittivity (δ)	3	10	6.6	9	9 [6]
Effective density of state in CB (cm^{-3})	2.5×10^{18}	2.25×10^{18}	1.2×10^{19}	2.2×10^{18}	2×10^{18}
Effective density of state in VB (cm^{-3})	1.8×10^{20}	10^{18}	2.9×10^{18}	1.8×10^{20}	1.8×10^{20}
Electron thermal velocity (cm/s^{-1})	10^7	10^7	10^7	10^7	10^7
Hole thermal velocity (cm/s^{-1})	10^7	10^7	10^7	10^7	10^7
Electron mobility ($cm^2/V\text{-}s$)	2×10^{-4}	2.2	2.7[3]	100	20
Hole mobility ($cm^2/V\text{-}s$)	2×10^{-4}	2.2	1.8 [3]	25	10
Shallow uniform acceptor density (cm^{-3})	10^{20}	10^{16}	10^{16}	0	0
Shallow uniform donor density (cm^{-3})	0	0	0	10^{19}	2×10^{19}
Holes capture cross section area (cm^2)	10^{-15}	10^{-13}	10^{-15}	10^{-15}	10^{-15}
Electrons capture cross section area (cm^2)	10^{-15}	10^{-13}	10^{-15}	10^{-15}	10^{-15}
Total Defect density N_t (cm^{-3})	10^{14}	10^{13}	10^{13}	10^{14}	10^{14} [6]

than $MAPbI_3$ has proven $FAPbI_3$ to be an attractive choice and appropriate alternative of $MAPbI_3$ absorbers in PSCs, indicating better thermal stability of $FAPbI_3$ perovskite film [4].

For the validation purpose the defect density, doping concentration, and thickness of $MAPbI_3$, $FAPbI_3$, ETL/HTL, and FTO, were kept constant. The defect energy level is characterized by neutral defect type with 0.6 eV energy level with respect to reference and single-level energy distribution (Fig. 22.2).

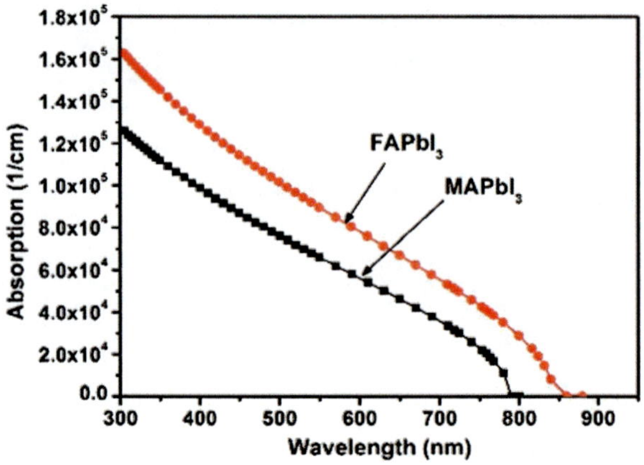

Fig. 22.1 SCAPS generated absorption versus wavelength plot for $MAPbI_3$ and FAPbI absorber materials

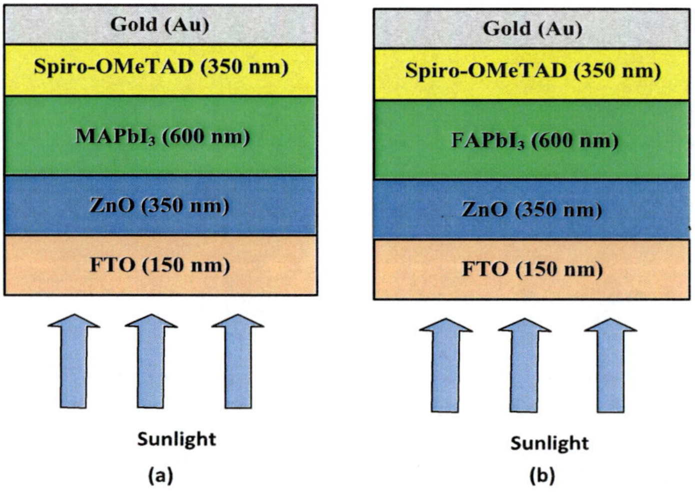

Fig. 22.2 Architectures of simulated **a** $MAPbI_3$ and **b** $FAPbI_3$ based PSCs in SCAPS

22.3 Results and Discussion

The results were recorded and analyzed in three phases. In the first phase, the performance of cells was compared based on the parameters listed in Table 22.1. The second and third phase of comparison was done based on defect density and acceptor density variations in terms of bandgap alignment, J-V characteristic, QE-λ characteristics, and recombination rate profiles.

22.3.1 Performance Comparison of $MAPbI_3$ and $FAPbI_3$ Cells

The permissible limit for the CBO and VBO are 0.2 and 0.3 eV respectively [8]. When VBO and CBO are more than 0.2 and 0.3 eV respectively, the interface recombination between HTL/perovskite and perovskite/ETL increase and performance of the cells degrade. As shown in Fig. 22.3a, in the case of $MAPbI_3$ absorber, the conduction band offset is 0.4 eV which is higher than the permissible limit (0.2 eV). The conduction band offset of $FAPbI_3$ is 0.29 eV which means the recombination rate at the perovskite/ETL interface is higher in $MAPbI_3$ based cell than $FAPbI_3$ cell as shown in Fig. 22.3d. The maximum recombination rate is observed in the case of $MAPbI_3$ is $4 \times 10^{19} cm^{-3}s^{-1}$ while $FAPbI_3$ has lower recombination rate as compared to $MAPbI_3$ that is $3.5 \times 10^{18} cm^{-3}s^{-1}$ from location 0.35 to 0.95 μm. The J-V characteristic shown in Fig. 22.3b in which $MAPbI_3$ shows the poor current density of 23.75 mA/cm^2 as compared to that in $FAPbI_3$ cell equal to 26.75 mA/cm^2. As shown in the result 3(c), $MAPbI_3$ material absorbs the photons in the UV–vis region from 300 to 750 nm of wavelength. While $FAPbI_3$ based solar cell can absorb the UV–vis plus part of IR spectra from 300 to 850 nm of wavelength.

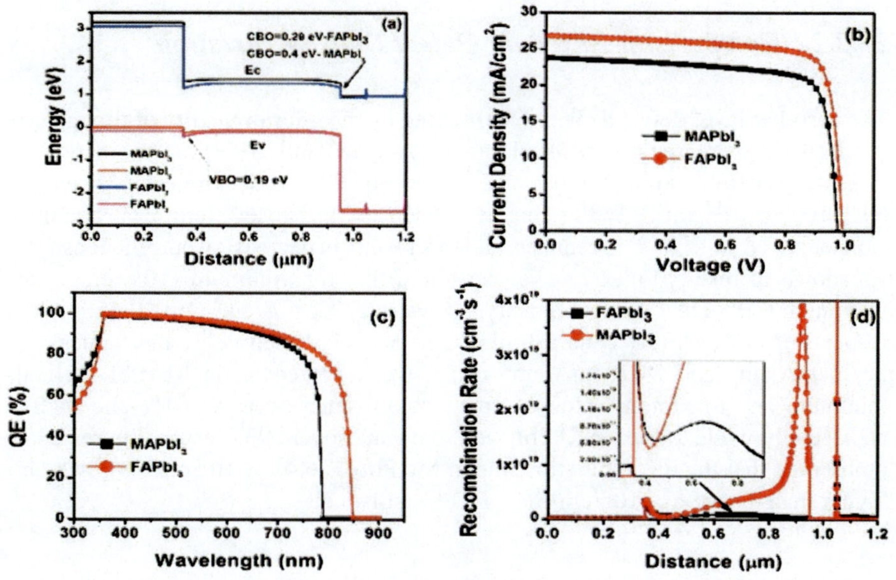

Fig. 22.3 **a** Energy band diagram, **b** J-V characteristic, **c** QE versus λ curve and **d** Recombination rate for $MAPbI_3$ and $FAPbI_3$

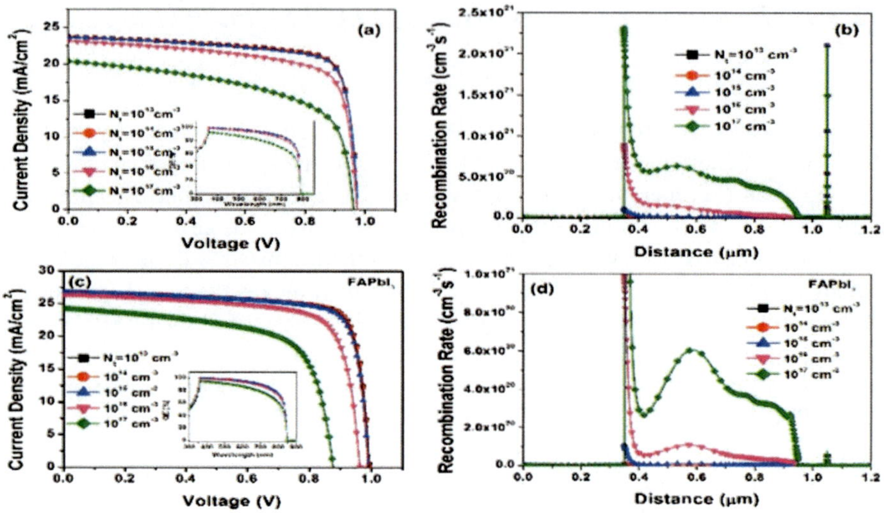

Fig. 22.4 **a** J-V characteristic **b** Recombination Rate, for $MAPbI_3$ based solar cell and **c** and **d**, shows the J-V characteristic and recombination rate for $FAPbI_3$ solar cell respectively, under defect density variation

22.3.2 Comparison Based on Defect Density Variation

The absorber layer defect density (N_t), caused by the inferior quality of film growth, results in the increased trap assisted Shockley–Read–Hall (SRH) recombination. An increased recombination rate was observed in Fig. 22.4 as defect density rises from $N_t = 10^{13}$ to 10^{17} cm^{-3}. With rising defect density the carrier lifetime (τ) and diffusion length ($L = \sqrt{D \times \tau}$) reduces, and this results in decreasing current density[9]. As shown in results Fig. 22.4a, c, we found that for minimum (10^{13} cm^{-3}) and maximum (10^{17} cm^{-3}) defect density, J_{SC} were 23.81 mA/cm^2 and 20.45 mA/cm^2 respectively in $MAPbI_3$ cells. Similarly, in $FAPbI_3$ based cells, the values were 26.76 mA/cm^2 and 24.28 mA/cm^2 respectively. Moreover, in $MAPbI_3$ cell, the minimum and maximum value of quantum efficiency were 98.81% and 99.27% respectively while those for $FAPbI_3$ cell, the values were 93% each. Figures 22.4b, d observed that the recombination rate in $MAPbI_3$ is $6.43 \times 10^{20} cm^{-3} s^{-1}$ which is higher than that of $6 \times 10^{20} cm^{-3} s^{-1}$ $FAPbI_3$ solar cell.

22.3.3 Comparison Based on Doping Density Variation

As shown in Fig. 22.5a, c, in $FAPbI_3$ and $MAPbI_3$ PSCs, the lower values of acceptor density variation (10^{13} to 10^{15} cm^{-3}), J_{SC} remains nearly equal to 26.5 and 23.3 mA/cm^2 while, for higher values (10^{16} to 10^{17} cm^{-3}), J_{SC} significantly

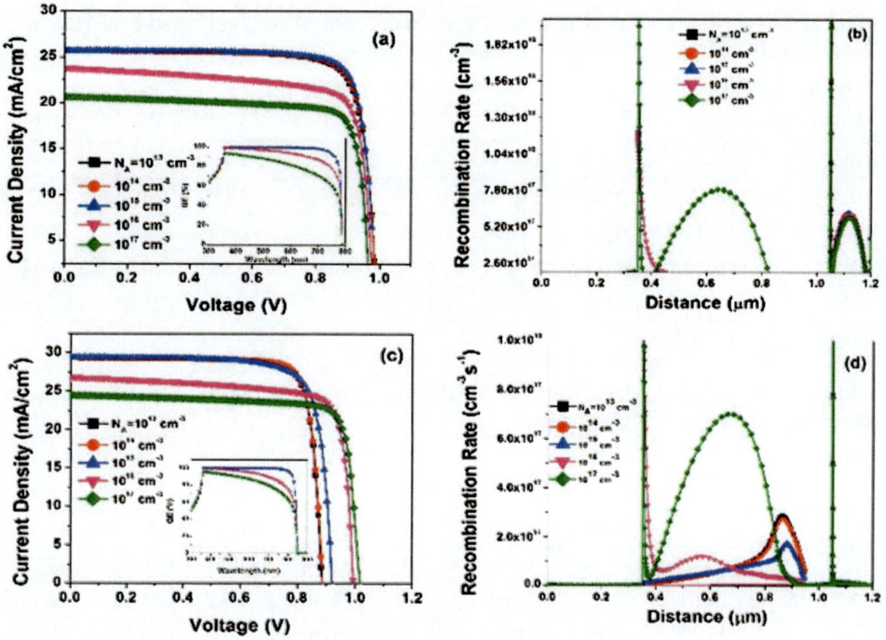

Fig. 22.5 **a** J-V characteristic **b** Recombination Rate, for $MAPbI_3$ based solar cell and **c** and **d**, shows the J-V characteristic and recombination rate for $FAPbI_3$ solar cell respectively, under Acceptor doping density variation

reduced to 23.76 and 20.52 mA/cm^2 as the recombination rate increase due to the rise of hole concentration respectively. In Fig. 22.5c, $FAPbI_3$ based solar cell V_{oc} decreases as N_A increases [10]. Here, $N_D = 0$, and N_A were varied to analyze the effect of N_A on V_{oc}. As the dark current is inversely proportional to V_{oc}, decrement in dark current causes the increment in V_{oc}. Figures 22.5b, d, shows that, as the defect or acceptor density increases, the photo-generated electrons and hole pairs recombine at a higher rate in the $MAPbI_3$ perovskite film as compared to $FAPbI_3$, and in the case of $MAPbI_3$ based perovskite solar cell the recombination rate found to be $8 \times 10^{17} cm^{-3} s^{-1}$ higher than $FAPbI_3$ solar cell equal to $6.97 \times 10^{17} cm^{-3} s^{-1}$.

Tables 22.2 and 22.3 demonstrate the performance parameters of the $MAPbI_3$ and $FAPbI_3$ based perovskite solar cell obtained from the different values of bulk defect density (N_t) and acceptor doping density (N_D). Thus we have found that, at the optimal value of $Nt = N_D = 10^{13}$ cm^{-3}, the power conversion efficiency is 21.68% for the $FAPbI_3$ and 20.63% PCE for $MAPbI_3$ based perovskite solar cell. The simulation results were validated by experimental data shown in Table 22.4.

Table 22.2 Performance Parameters for $MAPbI_3$ and $FAPbI_3$ based PSCs under the bulk defect density variation

$MAPbI_3$					$FAPbI_3$			
N_t (cm^{-3})	PCE %	V_{OC}	J_{SC}	FF %	PCE %	V_{OC}	J_{SC}	FF %
10^{13}	18.17	0.98	23.76	78.17	21.26	1.00	26.75	79.80
10^{14}	18.15	0.98	23.75	78.10	21.22	1.00	26.75	79.70
10^{15}	17.94	0.98	23.69	77.45	20.87	0.99	26.72	78.79
10^{16}	16.39	0.97	23.13	72.74	18.88	0.96	26.39	74.14
10^{17}	11.68	0.96	20.37	59.62	14.09	0.88	24.24	66.19

Table 22.3 Performance Parameters for $MAPbI_3$ and $FAPbI_3$ based PSCs under the absorber acceptor doping density (N_A) variation

$MAPbI_3$					$FAPbI_3$			
N_A (cm^{-3})	PCE %	V_{OC}	J_{SC}	FF %	PCE %	V_{OC}	J_{SC}	FF %
10^{13}	20.63	0.99	25.66	81.17	21.68	0.89	29.34	83.27
10^{14}	20.78	0.99	25.66	81.80	21.74	0.89	29.35	83.32
10^{15}	21.01	0.99	25.67	82.83	21.50	0.92	29.38	79.34
10^{16}	18.17	0.98	23.76	78.17	21.26	1.00	26.75	79.80
10^{17}	16.41	0.97	20.65	82.04	20.63	1.02	24.46	82.74

Table 22.4 Comparision of experimental and simulated results reported based on performance parameters of the solar cell as follows,

Structure		Voc (V)	Jsc in mA/cm^2	FF(%)	PCE (%)	Refs.
$MAPbI_3$/ZnO	Experimental	1.0	17.4	67	11.4	[11]
	Present Simulation	0.98	23.75	78.17	18.17	
$FAPbI_3$/ZnO	Experimental	1.07	20.5	68.5	15.0	[12]
	Present Simulation	0.99	26.75	79.8	21.26	

22.4 Conclusion

In summary, the comparative analysis was performed for the solution-processed $MAPbI_3$ and $FAPbI_3$ based solar cells with their attributes. Formamidinium based perovskite solar cell shows higher PCE of 21.26 % than 18.17% PCE of methylammonium based devices. Also, it has been observed that result shows a high conduction band offset $\Delta E_c = 0.4$ eV at the ETL/absorber interface in $MAPbI_3$ causes a rise in interfacial defect density whereas, in $FAPbI_3$ PSC, a lower value of conduction band offset $\Delta E_v = 0.19$ eV) observed at the ETL/absorber interface. Degradation of the performance of the $MAPbI_3$ solar cell was observed due to the higher value of CBO. J-V characteristic and Quantum efficiency characteristics are analyzed under the defect density and acceptor doping density variation. It is found that due to trap-assisted recombination, the current density decrease in both types of the solar cell as the defect density and acceptor doping density increase. Furthermore, the $FAPBI_3$ solar cell has higher stability and better light absorption ability than MA-based perovskite solar cells.

Acknowledgements Author Kshitij Bhargava is grateful to Dr. Marc Burgelman, Department of Electronics and Information Systems (ELIS), University of Gent, Belgium, for providing the SCAPS-1D simulation software.

References

1. Y. Wu, W. Chen, G. Chen, L. Liu, Z. He, R. Liu, The impact of hybrid compositional film/structure on organic-inorganic perovskite solar cells. Nanomaterials **8**(6), 1–27 (2018)
2. P. Tonui, S. O. Oseni, G. Sharma, Q. Yan, G. Tessema Mola, Perovskites photovoltaic solar cells: an overview of current status. Renew. Sustain. Energy Rev **91**, 1025–1044 (2018)
3. K. B. Nine, M. N. H. Shazon, S. A. Mahmood, Performance evaluation and comparative analysis of a highly efficient FAPbI 3-based perovskite solar cell. JOSA B **37**(10) October (2020)
4. Q. Wei, W. Zi, Z. Yang, D. Yang, Photoelectric performance and stability comparison of $MAPbI_3$ and $FAPbI_3$ perovskite solar cells. Sol. Energy **174** (September), 933–939 (2018)
5. S. Karthick, S. Velumani, J. Bouclé, Experimental and SCAPS simulated formamidinium perovskite solar cells: a comparison of device performance. Sol. Energy **205** (March), 349–357 (2020)
6. M. Yue Jie Su, Peng Zhao, Zhenhua Lin, Jincheng Zhang, Jingjing Chang, Yue Hao, Optimizing the performance of CsPbI3-based perovskite solar cells via doping a ZnO electron transport layer coupled with interface engineering. Nano-Micro Lett. **11**, 1–14 (2019)
7. G. Haidari, Comparative 1D optoelectrical simulation of the perovskite solar cell. AIP Advances **9**, 085028 (2019)
8. F. Unlu, E. Jung, J. Haddad, A. Kulkarni, S. Oz, H. Choi, T. Fischer, S. Chakraborty, T. Kirchartz, S. Mathur, Understanding the interplay of stability and efficiency in A-site engineered lead halide perovskites. APL Materials **8**(7) (2020)
9. A. Husainat, W. Ali, P. Cofie, J. Attia, J. Fuller, A. Darwish, Simulation and analysis method of different back metals contact of CH_3 NH_3PbI_3 perovskite solar cell along with electron transport layer TiO_2 using MBMT-MAPLE/PLD. Am. J. Opt. Photonics **8**(1), 6–26 (2020)
10. R. Jani, K. Bhargava, Comparative investigation into effects of the interplay between absorber layer crystallinity and interfacial defect states on the performance of lead-based and tin-based perovskite solar cells, in Semiconductor Science and Technology, IOP Publishing, pp. 1–22, (2020)
11. J. Song, W. Hu, X.F. Wang, G. Chen, W. Tian, T. Miyasaka, $HC(NH_2)_2PbI_3$ as a thermally stable absorber for efficient ZnO-based perovskite solar cells. J. Mater. Chem. A **4**(21), 8435–8443 (2016)
12. J. Yang, B. D. Siempelkamp, M. Edoardo, A. Filippo De, L K Timothy, Origin of the thermal instability in $CH_3NH_3PbI_3$ thin films deposited on ZnO. Chem. Mater. **27**, 4229−4236 (2015)

Chapter 23
Electrical Conductivity for Quasiparticle Graphene-Like System

Tanmay Das, Debakeenandan Pradhan, Anita Tamang, Jayanta Dey, Sabyasachi Ghosh, and Sesha Vempati

Abstract We have explored relaxation time approximation based kinetic theory framework on the electrical conductivity for different cases like non-relativistic, relativistic, extreme relativistic on graphene-like many body systems. Due to the same order of Fermi velocity for non-relativistic and graphene-like systems, which are quite lower than relativistic and extreme relativistic cases, kinetic theory provide same order of electrical conductivity, normalized by the order of relaxation time scales for metallic and graphene systems. We have found that the theoretical values of those normalized conductivity are quite close to the experimental values, which encourage us to practice the electronic dissipative hydrodynamics with deeper theoretical tools.

23.1 Introduction

Graphene is a 2D layer of one-atom-thick sp^2-bonded carbon atoms which is well known for its extraordinary electronic transport properties on the basal plane [1]. Electronic properties of graphene at energies below a few electron volts is well

T. Das (✉) · D. Pradhan · A. Tamang · J. Dey · S. Ghosh · S. Vempati
Department of Physics, Indian Institute of Technology Bhilai, GEC Campus, Sejbahar, Raipur, Chhattisgarh 492015, India
e-mail: tanmayd@iitbhilai.ac.in

D. Pradhan
e-mail: debakeenandanp@iitbhilai.ac.in

A. Tamang
e-mail: anitat@iitbhilai.ac.in

J. Dey
e-mail: jayantad@iitbhilai.ac.in

S. Ghosh
e-mail: sabya@iitbhilai.ac.in

S. Vempati
e-mail: sesha@iitbhilai.ac.in

R. G. Nair et al. (eds.), *Proceedings of 28th National Conference on Condensed Matter Physics*, Springer Proceedings in Physics 269,
https://doi.org/10.1007/978-981-16-5407-7_23

described by Dirac model [2–4]. Many authors have studied these by using the phenomenological two-dimensional Drude model, the current–current correlation function in the random phase approximation, the Kubo response formalism, and the Boltzmann transport equation [2, 4–9]. Recent studies [10, 11] described the electronic *hydrodynamics* in graphene system, where relaxation time approximation (RTA) based kinetic theory is well applicable to calculate the transport coefficients like electrical conductivity as done for other branches of research e.g. relativistic nuclear matter [12]. Interestingly graphene depicts an unconventional hydrodynamics, which is neither Galilean- nor Lorentz-invariant due to massless behavior of electrons in non-relativistic domain. In fact, the actual graphene samples can be little deviated from that of an ideal expectation [13].

Present work is intended to explore the RTA based kinetic theory calculations of electrical conductivity for ideal graphene (G) systems and its location with respect to traditional non-relativistic (NR), relativistic (R) and extreme relativistic (ER). By sketching microscopic kinematics of one body energy–momentum or dispersion relations and macroscopic quantity like electrical conductivity of the many-body system, a comparative numerical understanding of graphene with respect to other limiting cases (NR, R and ER) has been explored here. By juxtaposing the experimental data, we have presented an overall understanding of micro and macro-dynamics in terms of order of magnitude. We believe that this simplified investigation is in favor of recent prescription of electronic hydrodynamics and lots of future scope in theoretical studies, see recent [10] and references therein.

In the following, we discuss the 'Formalism', where detailed steps of RTA calculation of electrical conductivity in general and different special cases are addressed. Their numerical estimations are sketched and described in the results and discussion section. At the end we have summarized our study.

23.2 Methodology

Here we will obtain the microscopic expression of electrical conductivity in the general framework of RTA based kinetic theory while considering the special cases in convention of natural units. Under an external electric field, we can assume a non-equilibrium distribution $f = f_0 + \delta f$, which is quite close to equilibrium distribution with a small deviation δf. Hence, we get a non-zero net current density (due to external electric field)

$$J_x = 2e \int \frac{d^3k}{(2\pi)^3} v_x \delta f, \tag{23.1}$$

where, velocity along x axis or any direction can be averaged out as $v_x^2 \approx v_y^2 \approx v_z^2 \approx \frac{v^2}{3}$ with $\vec{v} = v_x\hat{x} + v_y\hat{y} + v_z\hat{z}$. Now, different possibilities of energy (ω) momentum (k) or dispersion relation for non-relativistic (NR), relativistic (R), extreme relativistic

(ER) and graphene-like (G) particles with mass m can be written.

$$\omega = \frac{k^2}{2m}(for\mathrm{NR}), \sqrt{k^2 + m^2}(for\mathrm{R}), k(for\mathrm{ER}), v_F k\ (for\mathrm{G}) \tag{23.2}$$

where v_F is the Fermi velocity of electrons in the graphene system. From the above relations, the velocity of the particle can be obtained.

$$v = \frac{d\omega}{dk} = \frac{k}{m}(for\mathrm{NR}), \frac{k}{\omega}(for\mathrm{R}), 1\ (for\mathrm{ER}), v_F\ (for\mathrm{G}) \tag{23.3}$$

Except for the R case, the (23.3) for NR, ER and G systems can be written in a general form:

$$\begin{aligned} \omega &= ak^n \\ \vec{v} &= \frac{d\omega}{dk} = an\,\vec{k}^{\,n-1} \end{aligned} \tag{23.4}$$

where $a = \frac{1}{2m}$, $n = 2$ for NR; $a = 1$, n $= 1$ for ER and $a = v_F$, n $= 1$ for G cases Since the deviation (δf) to the f_0 is originated from the external electrical field E_x and the functional dependency can be obtained by using RTA of Boltzmann equation (BE):

$$\begin{aligned} e\vec{E} \cdot \vec{\nabla}_k f_0 &= \frac{-\delta f}{\tau_c} \\ \Rightarrow \delta f &= -\tau_c e\vec{E} \cdot \left(\frac{d\omega}{d\vec{k}}\right)\left[\frac{\partial f_0}{\partial \omega}\right] \\ &= \tau_c e E_x (v_x)[\beta f_0[1 - f_0]] \end{aligned} \tag{23.5}$$

Using (23.5) in (23.1) and then comparing with Ohm's law $J_x = \sigma_{xx} E_x$, we will get.

$$J_x = \left[2e^2\beta \int \frac{d^3k}{(2\pi)^3} v_x^2 \tau_c f_0(1 - f_0)\right] E_x \tag{23.6}$$

$$\Rightarrow \sigma_{xx} = 2e^2\beta \int \frac{d^3k}{(2\pi)^3}\left(\frac{v^2}{3}\right)\tau_c f_0(1 - f_0) \tag{23.7}$$

$$\Rightarrow \sigma_{xx} = 2e^2\beta \int \frac{d^3k}{(2\pi)^3}\tau_c\left(\frac{1}{3}\right)a^2 n^2 \vec{k}^{\,2n-2} f_0(1 - f_0) \tag{23.8}$$

By given the isotropic nature of the conductivity, the general form of the expression for the conductivity including special cases (NR, G, ER) are as follows:

$$\sigma_{xx} = \sigma_{yy} = \sigma_{zz} = \sigma = \begin{cases} 2e^2\beta \int \frac{d^3k}{(2\pi)^3} \tau_c \left(\frac{1}{3}\right) a^2 n^2 \vec{k}^{2n-2} f_0(1 - f_0) & (23.9) \\ 2e^2\beta \int \frac{d^3k}{(2\pi)^3} \tau_c \frac{k^2}{3m^2} f_0(1 - f_0) \text{ (forNR)} & (23.10) \\ 2e^2\beta \left(\frac{v_F^2}{3}\right) \int \frac{d^3k}{(2\pi)^3} \tau_c f_0(1 - f_0) \text{ (forG)} & (23.11) \\ 2e^2\beta \frac{1}{3} \int \frac{d^3k}{(2\pi)^3} \tau_c f_0(1 - f_0) \text{ (forER)} & (23.12) \end{cases}$$

23.3 Results and Discussion

We start this section by plotting the well-known energy–momentum (23.3) relations in log scales and shown in Fig. 23.1. Log scale basically helps us to visualize all curves for ER (black dotted line), R (yellow solid line), NR (red solid line) and G (violet and green solid lines) cases together.

Both energy and momentum are normalized by mass in natural units to make it dimensionless so that they can be applicable to various systems. One can notice that at a high momentum domain, R curve is merging to its massless limit, described by the ER curve. On the other hand, NR curve is crossing the limit and exposes its failure to provide correct estimation at high momentum range. Interestingly, for the case of graphene in the presence of possible interactions with the substrate [14] the curve G within the range of Fermi velocity in natural units $v_F = 0.002 - 0.006$ depict much lower slope than that of R or ER curves, which signifies the critical nature of graphene-like systems. In one aspect, it follows the proportionality relation between energy and momentum, effectively the electrons in graphene behave like

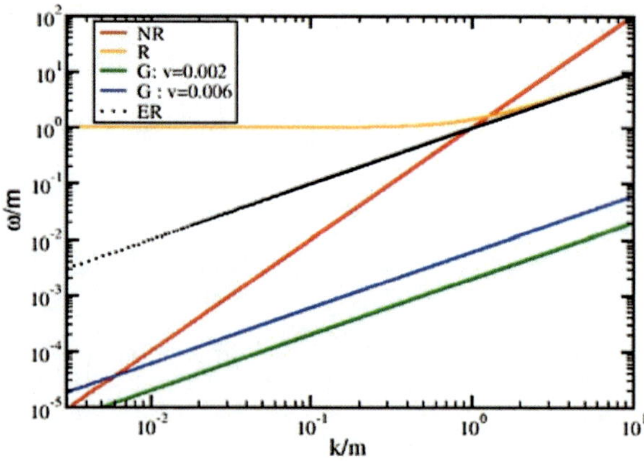

Fig. 23.1 Energy momentum (normalized) relation in natural unit (by making them both as dimensionless quantities and plotted in log scale) for different cases—ER (black dotted line), R (yellow), G (green and blue lines) NR (red)

massless particles. In other perspective, the behavior is far from the R and ER cases. This situation is never possible for free particles but a many-body environment can provide this kind of behavior to the particles, that is why the quasi-particle term is commonly used. Here, we have plotted energy–momentum curves with two of limiting values (0.002–0.006) corresponding to weak and strong electron–electron interactions [16]. With the help of energy–momentum relations, given in (23.3), one can obtain their corresponding velocity expression (4) and the square of which are plotted in Fig. 232a. From the RTA expression of conductivity, (8) it can be realized that its phase-space part as an average of square velocity followed by a normalization of relaxation time. Hence Fig. 23.2a also be considered as normalized conductivity as a function of momentum, which provides us with estimates of numerical strengths of different cases. However, in actual estimation of conductivity the momentum information is integrated by Fermi–Dirac distribution function. Figure 23.2b has provided that momentum integrated estimation.

For simplicity, we have considered degenerate electron gas (i.e. T = 0 eV). In the case of a metal with NR electron gas within Fermi energy range 2–10 eV the electrical conductivity is plotted in (23.3) (black solid line) normalized by the relaxation time and for comparison, experimental conductivity [15, 16] of silver (Ag), copper (Cu), gold (Au) and aluminum (Al) are shown by color circles. We have normalized the σ by suitable τ_c (values taken from [17]) and Fermi energy or chemical potential (values are taken from [18]). One can notice the close match between the estimated values from RTA and experimental values, which reflect the success of this traditional theory to describe electron transport in metals. Although we accept the fact that the match is in terms of numerical order. For simplicity, we have considered

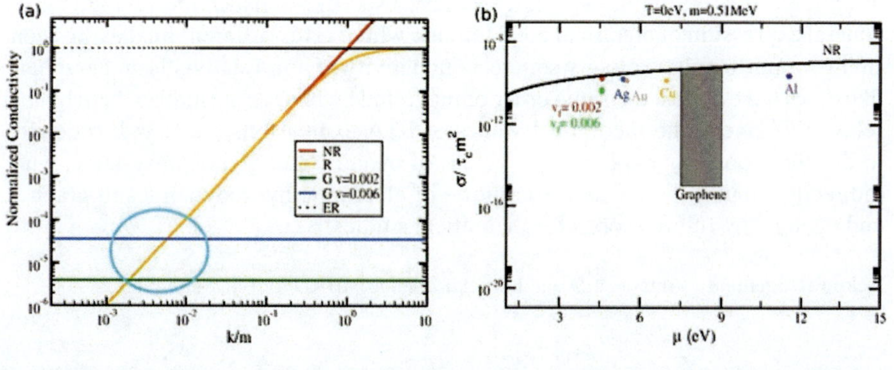

Fig. 23.2 **a** Roughly assuming conductivity as a proportional measurement of velocity square, normalized conductivity for ER (black dotted line), R (yellow), G (green and blue lines) NR (red) are plotted with momentum and **b** within 2–10 eV Fermi energy (typical range for metal), conductivity of non-relativistic (NR) electron gas is plotted by black solid line. For graphene-like (G) electron gas, red and green points stand for the corresponding values of conductivity at two Fermi velocity range $v_F = 0.002 - 0.006$. Corresponding experimental values [17, 18] of few metals (Ag, Au, Cu, Al) and a rough band of realistic graphene systems are included

step-function type FD distribution, which should be done for the room temperature distribution function. Now, when we go for the ideal graphene case, we have considered a Fermi energy of 4.6 eV and Fermi velocity range 0.002–0.006. Red and green points indicate the corresponding estimations. Their theoretical values are quite close to NR estimations, which is expected as we already noticed the same order of velocity in Fig. 23.2a. In kinetic theory, one can expect a similar projection from microscopic to macroscopic kinematics. When we go for experimental values of conductivity for different realistic graphene systems, we can get a range as shown in Gray rectangular box in Fig. 23.2b, constructed by combining the values available in [19–24]. We notice the RTA theory-based conductivity values of ideal graphene are within the band, made by experimental values of realistic graphene systems. Probably to explore lower experimental values, we might need other inputs like doping concentration, density of states etc. which will be incorporated in our future calculations.

23.4 Summary

In summary, we have gone through the microscopical calculations of electrical conductivity for free electron gas with different energy–momentum or dispersion relations like non-relativistic, relativistic, extreme relativistic and graphene-like relation. We have first sketched those quasi-particle kinematic relations to realize that electrons in graphene neither follow quadratic energy–momentum relation like non-relativistic case nor its velocity become close to the speed of light like relativistic or extreme relativistic cases. Next, we have used those dispersion relations in the general expression of electrical conductivity, which gives different final expressions. Being proportional to velocity square, conductivity of non-relativistic and graphene-like cases have almost the same order of magnitude when we normalize them by their relaxation times. The theoretical values of RTA expressions are found to be close to the corresponding experimental values of metallic and graphene systems, which indirectly support the recent prescription of electronic hydrodynamics in graphene and open many future scopes for theoretical studies.

Acknowledgements Authors T.D and J.D acknowledge MHRD for PhD fellowship.

References

1. K.S. Novoselov, A.K. Geim, S.V. Morozov, D. Jiang, Y. Zhang, S.V. Dubonos, Electric field effect in atomically thin carbon films. Science **306**, 666–669 (2004)
2. S. Droscher, F. Molitor, T. Ihn, K. Ensslin, in *Physics of Graphene* (Springer International Publishing, City, 2014), p. XII, 350

3. P.V. Buividovich, E.V. Luschevskaya, O.V. Pavlovsky, M.I. Polikarpov, M.V. Ulybyshev, Numerical study of the conductivity of graphene monolayer within the effective field theory approach. Phys. Rev. B **86**, 045107 (2012)
4. A.H.C. Neto, F. Guinea, N.M.R. Peres, K.S. Novoselov, A.K. Geim, The electronic properties of graphene. Rev. Mod. Phys. **81**, 109 (2009)
5. S.D. Sarma, S. Adam, E.H. Hwang, E. Rossi, Electronic transport in two-dimensional graphene. Rev. Mod. Phys **83**, 407 (2011)
6. K. Ziegler, Robust transport properties in graphene. Phys. Rev. Lett **97**, 266802 (2006)
7. K. Ziegler, Minimal conductivity of graphene: Nonuniversal values from the Kubo formula. Phys. Rev. B **75**, 233407 (2007)
8. M. Lewkowicz, B. Rosenstein, Dynamics of particle-hole pair creation in graphene. Phys. Rev. Lett **102**, 106802 (2009)
9. M.I. Katsnelson, *Graphene: Carbon in Two Dimensions* (Cambridge University Press, Cambridge, City, 2012), p. 363
10. B.N. Narozhny, Electronic hydrodynamics in grapheme. Ann. Phys. **411**, 167979 (2019)
11. B.N. Narozhny, Optical conductivity in graphene: hydrodynamic regime. Phys. Rev. B **100**, 115434 (2019)
12. C.A. Islam, J. Dey, S. Ghosh, Impact of different extended components of mean field models on transport coefficients of quark
13. T. Das, S. Vempati, Distinguishing strain, charge and molecular orbital induced effects on the electronic structure: graphene/ammonia system. J. Phys.: Condens. Matter **32**, 455501 (2020)
14. C. Hwang, D. Siegel, SK. Mo, Fermi velocity engineering in graphene by substrate modification. Sci Rep. **2**, 590 (2012)
15. R.A. Matula, Electrical resistivity of copper, gold, palladium, and silver. J. Phys. Chem. **8**, 1147–1298 (1979)
16. R.A. Serway, *Principles of Physics*, 2nd edn. (Fort Worth, Texas, London: Saunders College Pub, 1998), p. 602
17. D. Gall, Electron mean free path in elemental metals. J. Appl. Phys. **119**, 085101 (2016)
18. N.W. Ashcroft, N.D. Mermin, *Solid State Physics* (Saunders publisher, 1976)
19. S. Stankovich et al., Graphene-based composite materials. Nature **442**, 282–286 (2006)
20. M.A. Worsley et al., Synthesis of graphene aerogel with high electrical conductivity. J. Am. Chem. Soc. **132**(40), 14067–14069 (2010)
21. Z.H. Tang et al., Chem. Int. Ed. **49**, 4603 (2010)
22. Y. Xu, K. Sheng, C. Li, G. Shi, ACS Nano **4**, 4324 (2010)
23. Z.S. Wu et al., ACS Nano **3**, 411 (2009)
24. J. Xi, M. Long, L. Tang, D. Wanga, Z. Shuai, First-principles prediction of charge mobility in carbon and organic nanomaterials. Nanoscale **4**, 4348–4369 (2012)

Chapter 24
Quantum Hall Conductivity in Degenerate Electron Gas in Graphene-Like System

Debakeenandan Pradhan, Tanmay Das, Anita Tamang, Jayanta Dey, Sabyasachi Ghosh, and Sesha Vempati

Abstract Hall electrical conductivity is studied with relaxation time approximation based kinetic theory framework for metallic and graphene-like cases within a strong magnetic field-limit. A simple inversely proportional dependence on magnetic field in classical case is transformed to a complex field-dependent counterpart due to Landau quantization for quantum case. The order of magnitude of normalized conductivity is compared with the corresponding experimental values of graphene systems, which can be reduced to the order of e^2/h due to quantum Hall effect.

24.1 Introduction

Historically, Edwin Hall first introduce the classical Hall effect in 1879 [1], whose origin is basically (classical) Lorentz force, acting on charge particle in presence of magnetic field and then von Klitzing discovered quantum Hall effect (QHE) in 1980 [2], which disclose the quantum aspects of magnetic field, visible at low temperature and high magnetic field domain. Later in 2005, Novoselov et al. [3] have found the QHE in the graphene system, which can also be found in room temperature

D. Pradhan (✉) · T. Das · A. Tamang · J. Dey · S. Ghosh · S. Vempati
Indian Institute of Technology Bhilai, GEC Campus, Sejbahar, Raipur 492015, Chhattisgarh, India
e-mail: debakeenandanp@iitbhilai.ac.in

T. Das
e-mail: tanmayd@iitbhilai.ac.in

A. Tamang
e-mail: anitat@iitbhilai.ac.in

J. Dey
e-mail: jayantad@iitbhilai.ac.in

S. Ghosh
e-mail: sabya@iitbhilai.ac.in

S. Vempati
e-mail: sesha@iitbhilai.ac.in

R. G. Nair et al. (eds.), *Proceedings of 28th National Conference on Condensed Matter Physics*, Springer Proceedings in Physics 269,
https://doi.org/10.1007/978-981-16-5407-7_24

environments [4]. With respect to the dispersion relation, electrons in graphene are very critical in nature as they satisfy two contradictory aspects. Essentially, they remain within a non-relativistic zone, however, the massless nature of the electron is expected in an extreme relativistic condition. Here we have used relaxation time approximation (RTA) based kinetic theory calculations of electrical conductivity in presence of magnetic field for non-relativistic (NR) and graphene (G) systems. For numerical results, we have focused on the Hall conductivity component at zero temperature and strong magnetic fields, which appears to be dissipationless in nature because this limiting expression does not carry any information on relaxation time. Assuming NR expression as a metallic system within their known Fermi energy or chemical potential range, we can get the magnitude of their magneto-thermodynamic phase-space in Hall transport, which is compared with the corresponding values of the ideal graphene system. We believe that our investigation supports the recent description of electronic hydrodynamics [5–7], its magnetic field-extension [8, 9] and its practice in the relativistic nuclear matter [10, 11]. This direction recently opened many opportunities for future theoretical research.

The article is organized as follows. We first addressed the formalism with two subsections for classical and quantum Hall conductivity. Then, we have plotted their functional behavior, which are discussed in the results section and at the end we have summarized our work.

24.2 Formalism

Here, we will briefly address the electronic Hall conductivity component for classical and quantum cases in two consecutive subsections, where electrons are considered in different quasi-particle conditions like non-relativistic (NR), graphene-like (G) cases. The magnetic field (B) is considered along the z-axis, so the Hall component will be xy direction [12].

24.2.1 The Hall Conductivity for the Graphene-Like System: Classical Scenario

For classical case the energy–momentum ($\omega - k$) relations in the natural units for NR and G cases are as follows.

$$\omega = \begin{cases} \frac{k^2}{2m} & \text{(for NR)} \\ kv & \text{(forG)} \end{cases} \tag{24.1}$$

With the help of the RTA method, the Hall component of electrical conductivity at finite temperature ($T = \beta^{-1}$) and strong field limit can be described as [10, 11],

$$\sigma_{xy} = -2e^2\beta \int \frac{d^3k}{(2\pi)^3}\tau_B v_y^2 f_0(1 \mp f_0) \tag{24.2}$$

Here $f_0 = 1/\{e^{\beta(\omega-\mu)} + 1\}$ represents Fermi Dirac (FD) distribution function at a finite temperature (T), μ is chemical potential, v_y is y-component of Fermi velocity of electrons in graphene, e is charge of electron, $\tau_B = \frac{m}{eB}$.

Now, for a degenerate electron gas the FD distribution function $f_0 = 1/\{e^{\beta(\omega-\mu)} + 1\}$ will be converted to the step function $f_0 = \theta(\mu - \omega)$ and its derivative with respect to energy gives us the Dirac delta function. Hence, the Hall conductivity expression for degenerate gas will be,

$$\sigma_{xy} = \begin{cases} \frac{-2e^2}{6\pi^2 m}\tau_B(2m\mu)^{3/2}(for NR) \\ \frac{-2e^2}{6\pi^2 v}\tau_B\mu^2 \quad (for G) \end{cases} \tag{24.3}$$

24.2.2 *The Hall Conductivity for the Graphene-Like System: Quantum Scenario*

One can obtain momentum ($k_\perp$) quantization in a direction perpendicular to the magnetic field by solving the (NR) Schrodinger's equation or (relativistic) Dirac equation. As a result, we obtain, $k_\perp = \sqrt{k_x^2 + k_y^2} = 2leB$, where $l = 0, 1, 2 \ldots$ are Landau quantum numbers. This fact is known as the Landau quantization. Hence, the energy momentum relations in natural units for various cases (NR, and G) are as follows.

$$\omega_l = \begin{cases} \frac{k_z^2}{2m} + \frac{leB}{m} \text{ (for NR)} \\ v_F\sqrt{k_z^2 + 2leB} \text{ (for G)} \end{cases} \tag{24.4}$$

The phase space $2\int d^3k$ will be modified to $\sum_{l=0}^{\infty}\alpha_l \frac{|e|B}{2\pi}\int_{-\infty}^{+\infty}\frac{dk_z}{2\pi}$, where the spin degeneracy 2 will be converted to $\alpha_l = 2\ w\delta_{0,l}$ which will be 1 for lowest Landau level and 2 for remaining values of l. Also, we can assume $k_x^2 \approx k_y^2 \approx \frac{k_x^2+k_y^2}{2} = \frac{2l\tilde{e}B}{2}$. Adopting the above impositions, (24.2) becomes [10, 11],

$$\sigma_{xy} = -e^2\beta\sum_{l=0}^{\infty}\alpha_l\frac{eB}{2\pi}\int_{-\infty}^{+\infty}\frac{dk_z}{2\pi}v_y^2\tau_B f_0(1 \mp f_0) \tag{24.5}$$

Hence, the Hall conductivity expression for degenerate electron gas would be,

$$\sigma_{xy} = \frac{-e^2}{m} \sum_{l}^{l_{max}} \alpha_l \frac{|e|B}{(2\pi)^2} \tau_B \frac{\mu}{\sqrt{2m\mu - 2l|e|B}} \; (for NR) \tag{24.6}$$

$$= \frac{-e^2}{3} \sum_{l}^{l_{max}} \alpha_l \frac{|e|B}{(2\pi)^2} \tau_B \frac{\mu}{\sqrt{\left(\frac{\mu}{v}\right)^2 - 2l|e|B}} \; (for G) \tag{24.7}$$

where

$$l_{max} = \frac{2m\mu}{2eB} (for NR), \; \frac{\mu^2}{2eBv^2} \; (for G) \tag{24.8}$$

24.3 Results and Discussion

In the following, we discuss the electrical conductivity (Hall component only) from the graphical analysis of NR and G systems which are applicable to metal and graphene-like matter, respectively. A comparative study of the classical and quantum zones applicable for the systems is also done here.

In Fig. 24.1a and b the Hall conductivity (HC) normalized to the electron mass is plotted with respect to the magnetic field in Tesla. The Fermi energy of metal (NR system) generally varies within 2–10 eV. In Fig. 24.1a black solid and red dotted lines are for two extreme possible values of Fermi energies in the metallic system. The decrease in the HC for classical case can be explained by the inversely proportional magnetic field in (24.3) which actually governed by cyclotron frequency (τ_B). Now, the quantum version of HC, given in (24.6) and (24.7), are deduced from

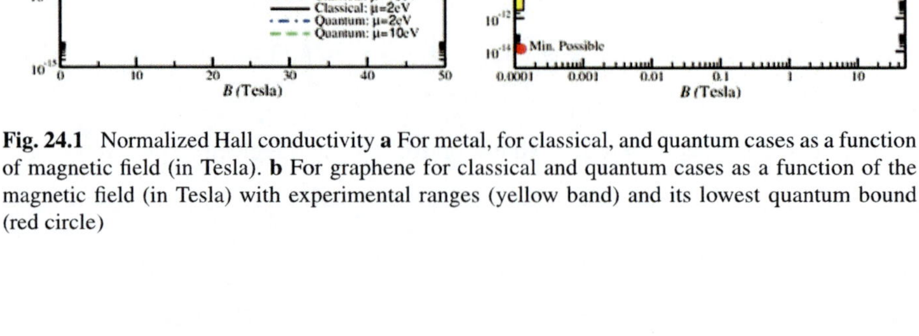

Fig. 24.1 Normalized Hall conductivity **a** For metal, for classical, and quantum cases as a function of magnetic field (in Tesla). **b** For graphene for classical and quantum cases as a function of the magnetic field (in Tesla) with experimental ranges (yellow band) and its lowest quantum bound (red circle)

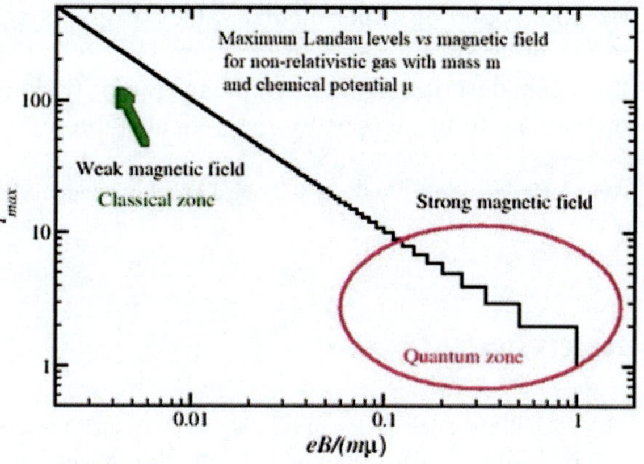

Fig. 24.2 Integer values of l_{max} versus eB/(mμ) where transition from classical to quantum regimes shown

classical results and little deviated from inverse relation due Landau quantization effect. Corresponding results for graphene-like systems are drawn in Fig. 24.1b, where we considered the speed of the electron (v_F) as a free parameter and the results are for two values of v_F (=0.002, 0.006) [13] given in terms of the speed of light. They also decrease similarly to NR cases. We should concentrate on the B = 1–50 T zone, which might be considered as a strong field limit for metallic and graphene systems at very low temperature (for calculation purposes it is considered as 0 K). Our normalized Hall conductivity curves basically represent the magneto-thermodynamic phase-space factor of Hall transport in a strong field limit. Experimental values of normalized conductivity for graphene systems fall within a range (marked by yellow band in Fig. 24.1b) by combining the values available in [14–19], which can be reduced to its lowest quantum bound e^2/h, marked by red circle in Fig. 24.1b.

The maximum value of Landau level is given by (24.8). It is plotted against $\frac{eB}{m\mu}$ for NR case in Fig. 24.2, from where a transition from continuous to discretization can be realized. Imposing this restriction, the pattern of continuous to discrete in conductivity components can be obtained. A detailed work to explain the conductivity, in particular, metallic and graphene-like systems under different extensions like finite temperature calculation, interaction between the particles is in progress.

24.4 Summary

In summary, we have addressed the microscopic calculations of Hall conductivity for metallic and graphene systems in presence of strong magnetic fields, which disclose a dissipationless magneto-thermodynamic transport. Their classical and quantum degenerate cases are obtained by considering without and with Landau quantization methods. A simple inversely proportional dependence on magnetic field in classical case is transformed to a complex field dependence due to Landau quantization for quantum case. The order of magnitude of normalized conductivity is compared with the corresponding experimental values of graphene systems, which can be reduced to the order of e^2/h due to quantum Hall effect. Present work indirectly support the recent description of electronic magneto-hydrodynamics in graphene, which will open many future scopes for theoretical studies.

Acknowledgements Authors T.D and J.D acknowledge MHRD for PhD fellowships.

References

1. E. Hall, On a new action of the magnet on electric currents. Am. J. Math. **2**(3), 287–292 (1879)
2. K. v Klitzing, G. Dorda, M. Pepper, New method for high-accuracy determination of the fine-structure constant based on quantized hall resistance. Phys. Rev. Lett. **45**, 494 (1980)
3. K.S. Novoselov, A.K. Geim, S.V. Morozov, D. Jiang, M.I. Katsnelson, I.V. Grigorieva, S.V. Dubonos, A.A. Firsov, Two-dimensional gas of massless Dirac fermions in graphene. Nature **438**, 197–200 (2005)
4. K.S. Novoselov et al., Science **315**, 1379 (2007)
5. B.N. Narozhny, Electronic hydrodynamics in graphene, Ann. Phys. 411, 167979 (2019)
6. B.N. Narozhny, I.V. Gornyi, M. Titov, M. Schutt, A.D. Mirlin, Hydrodynamics in graphene: linear-response transport. Phys. Rev. B **91**, 035414 (2015)
7. A. Lucas, K. Chung Fong, Hydrodynamics of electrons in grapheme. J. Phys.: Condens. Matter **30**, 053001 (2018)
8. B.N. Narozhny, M. Schutt, Magnetohydrodynamics in graphene: shear and Hall viscosities. Phys. Rev. B **100**, 035125 (2019)
9. M. Muller, L. Fritz, S. Sachdev, Quantum-critical relativistic magnetotransport in grapheme. Phys. Rev. B **78**, 115406 (2008).
10. A. Bandyopadhyay, S. Ghosh, R.L.S. Farias, J. Dey, G. Krein, Anisotropic electrical conductivity of magnetized hot quark matter. Phys. Rev. D **102**, 114015 (2020)
11. S. Samanta, J. Dey, S. Satapathy, S. Ghosh, Quantum expression of electrical conductivity from massless quark matter to hadron resonance gas in presence of magnetic field (2020), arXiv:2002.04434
12. D. Tong, Lectures on the quantum hall effect, arXiv:1606.06687 [hep-th]
13. C. Hwang, D. Siegel, SK. Mo, Fermi velocity engineering in graphene by substrate modification. Sci Rep. **2**, 590 (2012)
14. S. Stankovich et al., Graphene-based composite materials. Nature **442**, 282–286 (2006)
15. M.A. Worsley et al., Synthesis of graphene aerogel with high electrical conductivity. J. Am. Chem. Soc. **132**(40), 14067–14069 (2010)
16. Z.H. Tang et al., Chem. Int. Ed. **49**, 4603 (2010)
17. Y. Xu, K. Sheng, C. Li, G. Shi, ACS Nano **4**, 4324 (2010)
18. Z.S. Wu et al., ACS Nano **3**, 411 (2009)
19. J. Xi, M. Long, L. Tang, D. Wanga, Z. Shuai, First-principles prediction of charge mobility in carbon and organic nanomaterials. Nanoscale **4**, 4348–4369 (2012)

Chapter 25
Recent Advances in Magnetically Separable g-C_3N_4 Based Multi-component Nanocomposites for Visible-Light Driven Photo-Catalysis

Suma Das and Avijit Chowdhury

Abstract The graphitic carbon nitride (g-C_3N_4) has received tremendous attention for its robust, economical, and excellent light-induced properties to counter various scientific challenges, including environmental pollution mitigation. However, many factors such as high electron–hole recombination rate, lesser surface area, poor electrical conductivity, and recyclability of the bulk g-C_3N_4 eventually lower the photo-catalytic efficiency and limit their reusability and practical application. The most effective approach to tackle the aforementioned problem is the exploitation and integration of the magnetic and semiconducting materials with g-C_3N_4. The hetero-junction photo-catalysts are expected to improve photo-catalytic efficiency through a synergistic interfacial effect. Further, the magnetically separable nanoparticles (NPs) grafted onto the surface of photo-catalysts ease the effective separation process of the catalysts without any difficulties. This paper is an updated and all-inclusive review that briefly discusses the photo-catalytic mechanism of g-C_3N_4-based multi-compound nanocomposites (NCs) consisting of ternary and quaternary semiconductor. Moreover, the effects of semiconductor (SC) hetero-junctions on the structural, physiochemical, and the photo-catalytic performances are also discussed in details.

25.1 Introduction

The band-gap engineering of the semiconducting materials is in the hotspot of materials science to explore in the visible-light-induced photo-catalytic processes including their potential applications in many disciplines [1]. Apart from the convenience aspects, SC-based photo-catalyst suffers from several drawbacks which restricts their practical impediments that need to be addressed [2]. The most widely used SC photo-catalysts are photoactive only in the ultraviolet (UV) region of the electromagnetic spectrum due to relatively wider energy band gaps (BGs) [3]. It

S. Das (✉) · A. Chowdhury
Organic Electronics & Sensor Laboratory, Department of Physics, National Institute of Technology Silchar, Cachar, Assam 788010, India

R. G. Nair et al. (eds.), *Proceedings of 28th National Conference on Condensed Matter Physics*, Springer Proceedings in Physics 269,
https://doi.org/10.1007/978-981-16-5407-7_25

is evident from the spectral distribution of the solar irradiation that suitable band-gap materials along with materials engineering could be the only viable solutions to address the major issues. Till date, many research efforts are devoted to prepare suitable photo active catalysts but still the major challenges are lying mainly on their activity and sustainability. The metal free polymeric n-type SC, g-C_3N_4 are explored in the photo-catalytic processes due to their tunable energy BG (~2.7 eV), 2D layered structure, high chemical and photochemical stability, and low cost [4]. However, the pristine g-C_3N_4 displays poor photo-catalytic performances due to high recombination rate, smaller surface area [5] and lower electrical conductivity [6]. Another issue is the recyclability of the g-C_3N_4-based photo-catalysts because of their highly dispersive nature which further restricts their practical applications.

As per the literature, many separation methods are adopted till date to recover the g-C_3N_4-based photo-catalysts. However, it is observed that the lack of facile and environment friendly technique ends up with the loss of catalysts [7]. The magnetic photo-catalysts, comprising of magnetic and photoactive materials, can be isolated effectively from their suspension using an external magnetic field [8]. This novel method opens up a new avenue which leads to low-cost, low time consumption and better reusability property of the nanosized photo-catalysts. Recently, IONPs has taken the whole spotlight in which the magnetic photo-catalysts are prepared by following the cost-effective method and using the abundant materials in order to remove organic pollutants [9]. However, there are still challenges to developed highly visible-light driven photo-catalyst material to hinder fast e^--h^+ recombination in order to improve the photo-catalytic performances. A number of approaches are becoming popular for the development of novel SC-based photo-catalyst, such as combining two or more SCs through BG engineering, fabrication of NCs, tuning of physical properties, doping and creating active sites. The most effective approach to tackle the aforesaid problem is to exploitation and integration of the magnetic as well as semiconducting materials with g-C_3N_4 [10].

25.1.1 Magnetic Photo-Catalyst Based on g-C_3N_4

The different phases of IONPs are investigated based on their characteristics and attributes, such as Fe_3O_4, α-Fe_2O_3, β-Fe_2O_3, γ-Fe_2O_3, MFe_2O_4 (M = Mn, Fe, Zn, Ni, Co) [11]. Among these, Magnetite (Fe_3O_4) NPs are thoroughly used in various applications [12]. This material possesses several advantages like large surface area, low toxicity [13], low-cost [14], simple preparation method, and, most importantly, their super-paramagnetic and high saturation magnetization [15]. However, the Fe_3O_4 NPs has the tendency to deform and agglomerate during the catalyst recovery process which significantly hinders their large-scale application [13]. Therefore, grafting of g-C_3N_4 nanosheets with Fe_3O_4 NPs is attempted in various studies to enhance the photo-catalytic performance. This performance improvement is understood through the matching energy band positions in the Fe_3O_4/g-C_3N_4 NCs. Upon white light illumination, the photo-excited electrons can easily transfer from the CB of g-C_3N_4

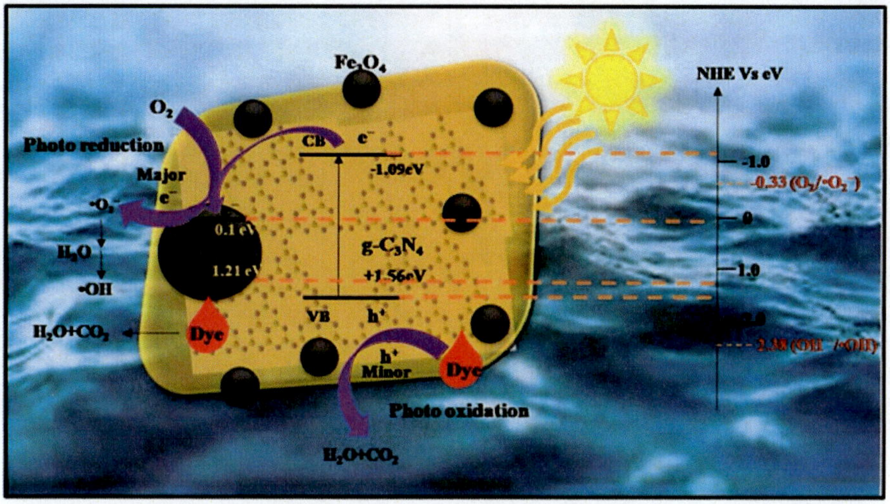

Fig. 25.1 Schematic illustration of the mechanism for photo-induced charge carrier transfers in the g-C_3N_4/Fe_3O_4 photo-catalyst under visible-light irradiation

to the CB of Fe_3O_4 since the CB of Fe_3O_4 is slightly lower than that of g-C_3N_4. However, the charge transport in Fe_3O_4 NPs is comparatively fast due to the narrow BG and high electrical conductivity [16]. Consequently, the probability of the direct recombination of photo-induced carriers is minimized by creating large number of Reactive Organic Species (ROS) in the Fe_3O_4/g-C_3N_4 NCs [17]. The electron in the CB of Fe_3O_4 assists efficient reduction of absorbed O_2 into different ROS to produce water, CO_2 etc. through the generation of highly ROS-OH (as shown in Fig. 25.1).

In a report, Kumar et al. [12] have successfully prepared g-C_3N_4/Fe_3O_4 NCs via a cost-effective and simple in-situ growth mechanism wherein Jia et al. have synthesized the same NC via a hydrothermal route [13]. In both the cases, g-C_3N_4 nanosheets are decorated by Fe_3O_4NPs with diameter size of about 6–8 nm that efficiently prevented agglomeration of Fe_3O_4NPs. Interestingly, compared to in-situ growth, a large specific surface area about 200.83 and 115.1 m^2/g of g-C_3N_4/Fe_3O_4 is obtained via hydrothermal and calcinations-co-precipitation method [16] Using these materials, a significant enhancement in the photo-catalytic performances for the degradation of Rhodamine B (RhB) is observed. This improvement in the photo-catalytic activity is documented as the synergistic interfacial effect between Fe_3O_4 and g-C_3N_4 [13]. On the other hand, Liu et al. have reported an efficient performance of hydrothermally prepared Fe_3O_4/g-C_3N_4 NC for the degradation of RhB using H_2O_2 as an oxidant [17]. In addition, ascribed to the super-paramagnetic behavior, g-C_3N_4/Fe_3O_4 NCs are recovered and reused even after 5–6 successive cycles without any loss of photo-catalytic activity. Therefore, different preparation routes effects on several properties of as prepared NCs which are tabulated below in Table 25.1.

Even though, the use of Fe_3O_4 NPs grafted g-C_3N_4 surface resolves the problem of easy separation and recycling of non-magnetic catalyst but the industrial application

Table 25.1 Photo-catalytic performances of Fe_3O_4/g-C_3N_4-based magnetic photo-catalyst

Photo-catalyst	Preparation method	M_s (emu/g)	Pollutant	Light source	Degradation rate/time (min)	Run	References
Fe_3O_4/g-C_3N_4	In situ growth	–	RhB	300 W Xe lamp	–	6	Kumar et al. [12]
Fe_3O_4/g-C_3N_4	Hydrothermal	47.6	RhB	500 W Xe lamp	95.5%/60 min	5	Jia et al. [13]
Fe_3O_4/g-C_3N_4	Calcinations& Co-precipitation	80.49	2,4,6trichlorophenol	500 W Xe lamp	95.7%/100 min	5	Yang et al. [16]
Flower-like-flake Fe_3O_4/g-C_3N_4	In situ growth	–	RhB	300 W Xe lamp	97.83%/140 min	5	Zhu et al. [14]
Fe_3O_4@ g-C_3N_4 + H_2 O_2	Hydrothermal	5	RhB MB	Visible light	100%/120 min 100%/180 min	4	Liu et al. [17]

of g-C_3N_4 is still restricted. Currently, numerous SCs with suitable band edge positions are combined to enhance the light absorption abilities as well as to prolong life time of photo-generated EHP. In light of this, g-C_3N_4-based ternary and quaternary NCs have engrossed extensive attention by the researchers in order to get the collective advantages of employed components through the formation of hetero-junction between constituents. Based on band edge positions, the hybrid NCs are further subdivided into following four categories that are g-C_3N_4/Fe_3O_4/narrow band gap SC, g-C_3N_4/Fe_3O_4/wide band gap SC, g-C_3N_4/Fe_3O_4/narrow/narrow band gap SC and g-C_3N_4 /Fe_3O_4/narrow/wide band gap SC.

25.1.1.1 Ternary Nanocomposites (NCs)

In order to form hetero-junction structure to stimulate photo-activity, various narrow and wide BG SCs have been integrated with g-C_3N_4 [18]. Coupling of narrow BG SC with g-C_3N_4, synergistically enhances the photon absorption ability as compared to pure g-C_3N_4. Whereas, combining wide BG SC with proper band edge, suppresses the rate of photo-generated e^--h^+ recombination, which is well explained by producing and characterizing several magnetically separable ternary NCs by the researchers. In recent years, coupling of noble metal NPs onto SC surface has gained great interest in the field of pollutant degradation harnessing the local Surface Plasmon Resonance effect (SPR). For example, Zhu et al. have prepared highly dispersed Ag/ Fe_3O_4/g-C_3N_4 composites which improve the photo-catalytic activity by retaining the magnetic properties as well. This improvement is ascribed as the electrons trapping within Fe_3O_4 and then transfer to Ag. This mechanism leads to higher separation efficiency of EHP and hence more light harvest [18]. Further, a series of Fe_3O_4/g-C_3N_4/Ag_3VO_4 and Fe_3O_4/ g-C_3N_4/Ag_2CrO_4 are synthesized via refluxing method [19, 20]. Both the NCs shows a decrease in saturation magnetization compared to Fe_3O_4, which is yet enough to separate it from the treated solution using magnet. Compared to the absorption edge of pure g-C_3N_4, the NCs Fe_3O_4/g-C_3N_4/Ag_3VO_4 (60%) and Fe_3O_4/g-C_3N_4/Ag_2CrO_4(20%) exhibit strong absorption in the visible region, that revealed superior photo-catalytic activities for degradation of Rh B. This enhancement in the activity was assigned to the effective separation of charge carriers through transfer of photo-generated e^- and h^+ s as shown in Fig. 25.2.

Since the E_g values for g-C_3N_4, Ag_3VO_4 and Ag_2CrO_4 are 2.7, 2.2 and 1.8 eV, hence under light irradiation both the counter parts of composites generate EHP. The CB and VB edge positions of g-C_3N_4 are more negative than Ag_3VO_4 or Ag_2CrO_4. As a result, photo-generated electrons can easily transfer from g-C_3N_4 toAg_3VO_4or Ag_2CrO_4 and holes in the reverse direction. Therefore, the effective separation of photo-generated charge carriers takes place in the composites resulting in lowering of e^--h^+ recombination rate. Likewise, a series of g-C_3N_4/Fe_3O_4/$CoWO_4$ (10%), and g-C_3N_4/Fe_3O_4/$NiWO_4$ (10%) NCs are recently synthesized using facile refluxing-calcinations method. An improved photo-catalytic performance of 12.5 and 12 times higher for Rh B; 27.5 and 30 times higher for MB; 45.7 and 52 times higher for MO and 102 and 100 times higher for fuchsine relative to g-C_3N_4 are achieved.

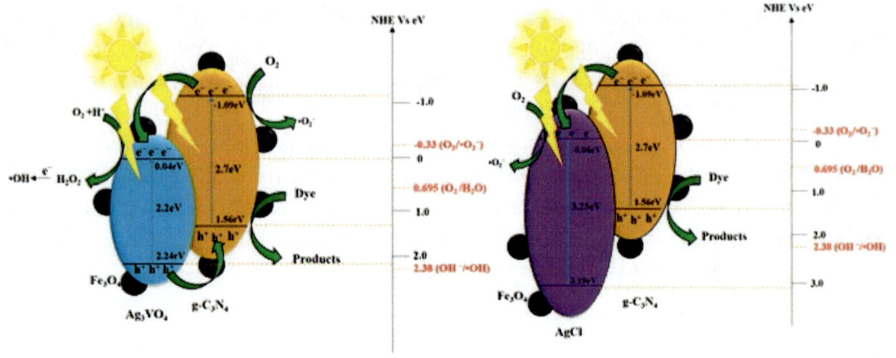

Fig. 25.2 Schematic illustration of proposed degradation mechanism of organic dye over the Fe_3O_4/g-C_3N_4/$Ag_3VO_4$and Fe_3O_4/ g-C_3N_4/AgCl nanocomposites

The improved performances of the ternary photo-catalysts are correlated with large specific surface area, and excellent ability for absorption of visible light. It is revealed that the superoxide anion radicals are the most ROS in the photo-degradation reaction [21, 22]. In another report, g-C_3N_4/Fe_3O_4/AgCl (40%) is prepared by a facile method [23], which shows 7.7 times increase in the photo-catalytic activity than pure g-C_3N_4. The maximum saturation magnetization of 9.5 emu is obtained per gram of the g-C_3N_4/Fe_3O_4/AgCl (40%) nanocomposite, describing its potential magnetic separation. Compared to Ag_3VO_4 and Ag_2CrO_4, the E_g value for AgCl (3.25 eV) is higher due to which under visible light irradiation the EHP are generated only in the g-C_3N_4 counterpart of the NC. As a result, photo-induced carriers are easily transferred leading to a decrease in recombination rate.

25.1.1.2 Quaternary Nanocomposites (NCs)

As we discussed above, ternary NCs composed of narrow BG SC gives wide absorption in the whole visible region where as NCs composed of wide BG SC effectively reduces the direct recombination of photogenerated EHP. Till now, several quaternary NCs are prepared in order to achieve both the goals simultaneously. The g-C_3N_4/Fe_3O_4/Ag_3VO_4/Ag_2CrO_4NCs are prepared by a facile refluxing method [24]. It is reported that since the CB potential of Ag_3VO_4(+0.04 eV) and Ag_2CrO_4 (+0.46 eV) are more negative than that of the redox potential O_2/H_2O_2 (+0.695 eV Vs NHE) therefore the CB electrons produces H_2O_2 molecules by reacting with absorbed oxygen molecules and further reacting with additional electrons these molecules consequently produces active hydroxyl radicals [24]. It demonstrated that the g-C_3N_4/Fe_3O_4/Ag_3VO_4/Ag_2CrO_4 NCs exposed significant enhancement in photo-degradation of Rh B, MB, and MO under white light irradiation as compared to their individual counterparts. The interfacial contact among g-C_3N_4, Ag_3VO_4 and Ag_2CrO_4 in the NCs greatly improves the separation of photo-induced EHP but

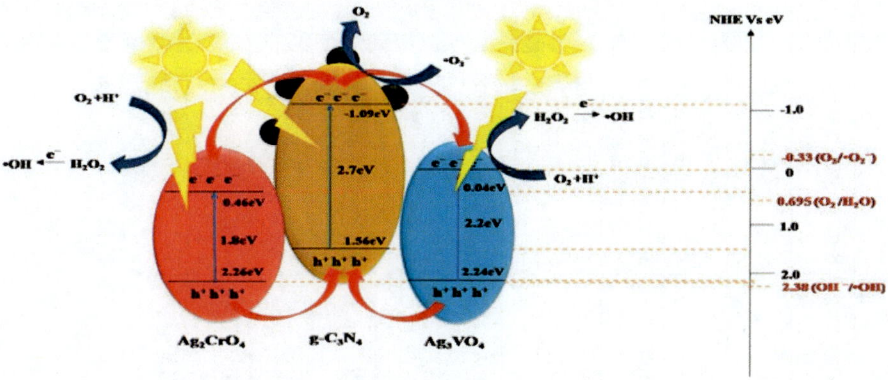

Fig. 25.3 Schematic illustration of proposed degradation mechanism of organic dye over the g-$C_3N_4/Fe_3O_4/Ag_3VO_4/Ag_2CrO_4$ quaternary NCs

reduces the saturation magnetization. Due to narrow BG of the SCs, all the counter parts of the composite produce photo-generated EHP under visible light irradiation as shown in Fig. 25.3.

In g-$C_3N_4/Fe_3O_4/Ag_2WO_4$/AgBr NCs, e^--h^+ are generated only in g-C_3N_4 and AgBr not in Ag_2WO_4 due its wide BG. The band edge positions of g-C_3N_4 are more negative than AgBr and Ag_2WO_4. So, the electrons can easily move from CB of g-C_3N_4 to AgBr as well as Ag_2WO_4. Therefore, the effective separation EHP leads to an enhanced photo-catalytic activity. The g-$C_3N_4/Fe_3O_4/Ag_2WO_4$/AgBr (30%) NC exhibited 94.7, 17.4, and 26.7-fold enhanced activity relative to g-C_3N_4, g-$C_3N_4/Fe_3O_4/Ag_2WO_4$(30%) and g-C_3N_4/Fe_3O_4/AgBr (30%), respectively for the degradation of MO [25]. Kumar et al. have prepared Fe_3O_4/BiOCl/g-C_3N_4/Cu_2O (p-n-p) hetero-junction with excellent photo-catalytic activity by facile co-precipitation method [26]. It exhibits superior photo-catalytic activity for degradation of Sulfamethoxazole (SME) with 99.5% in 60 min under visible light. On the formation of p-n-p hetero-junction, the electrons flow from BiOCl and Cu_2O to that of g-C_3N_4, this leads to rising of an inbuilt electric field that pushes the charge carriers to flow and reduces carrier recombination. Table 25.2 offers a summary of few fabricated ternary and quaternary NCs and their photo-catalytic performances for degradation of organic pollutants.

25.2 Conclusions and Perspectives

Despite of many appealing characteristics, the application of bulk g-C_3N_4 is restricted due to a number of inherent limitations in the field of water decontamination. The effective and simple separation of g-C_3N_4 from the treated suspension is considered as one of the most challenging tasks for reusability. The incorporation of g-C_3N_4 with magnetite materials are systematically discussed which comes out to be an

Table 25.2 Photo-catalytic performances of Fe_3O_4/ g-C_3N_4-based ternary and quaternary NCs

Photo-catalyst	Preparation method	M_s(emu/g)	Pollutant	Light source	Degradation rate/ time (min)	Run	References
Ag(3wt%) /Fe_3O_4/g-C_3N_4	Photo-deposition	12.68	Tetracycline (TC)	300 W Xe lamp	88%/90	–	Zhu et al. [18]
Fe_3O_4/ g-C_3N_4/Ag_3VO_4 (60%)	Refluxing	5.1	RhB	50 W LED	98%/240	5	Mousavi and Yangjeh [19]
Fe_3O_4/g-C_3N_4/AgCl (40%)	In situ growth	9.5	RhB	50 W LED	98.3%/360	5	Akhundi and Yangjeh [23]
Fe_3O_4/ g-C_3N_4/Ag_2CrO_4(20%)	Refluxing	12.9	RhB	50 W LED	95%/300	5	Yangjeh and Akhundi [20]
Fe_3O_4/g-C_3N_4/Bi_2S_3	Refluxing	8.12	RhB	50 W LED	100% 40	4	Musavi et al. [15]
Fe_3O_4/ g-C_3N_4/$CoWO_4$ (10%)	Refluxing-Calcination	7.04	RhB	50 W LED	100%/150	4	Mousavi and Yangjeh [21]
Fe_3O_4/ g-C_3N_4/$NiWO_4$ (10%)	Refluxing-Calcination	6	RhB	50 W LED	100%/240	4	Mousavi and Yangjeh [22]
Fe_3O_4/g-C_3N_4/TiO_2	Impregnation	0.1	MB	30 W Hg lamp	96%/210	–	Abbasia and Farrokhniaa [27]
Fe_3O_4/ g-C_3N_4/$CoMoO_4$(30%)	Refluxing-Calcination	20.1	RhB	50 W LED	~100%/300	4	[Yangjeh et al. 28]
Fe_3O_4/g-C_3N_4/MoO_3(30%)	Calcination	26.3	TC	1000 W Xe lamp	94%/120	–	Hel et al. [29]
Fe_3O_4/ g-C_3N_4/Ag_3VO_4/ Ag_2CrO_4 (20%)	Refluxing	6.12	RhB	50 W LED	~100%/180	–	Mousavi and Yangjeh [24]

(continued)

Table 25.2 (continued)

Photo-catalyst	Preparation method	M_s(emu/g)	Pollutant	Light source	Degradation rate/ time (min)	Run	References
Fe_3O_4/ g-C_3N_4/Ag_2WO_4/ AgBr(30%)	Refluxing	9.61	RhB	50 W LED	100%/150	4	Akhundi and Yangjeh [25]
Fe_3O_4/BiOCl/ g-C_3N_4/ Cu_2O	Co-precipitation	26	(SME)	800 W Xe lamp	99.5%/60	5	Kumara et al. [26]

effective strategy to overcome the aforesaid shortcoming. Alongside, the integration of SC material improves visible-light response, facilitates the photo-generated EHP, extended charge separation duration and subsequently boosts the photo-catalytic activity. The magnetic g-C_3N_4-based NCs are good enough to use in successive runs, however, further study is required to establish their visible-light absorption ability via suitable and cost-effective preparation method.

References

1. L. Cui, T. Pu, X. Fang, J. Song, S. Li, J. Wang, C. Yin, H. Shi, S. Kang, Int. J. Electrochem. Sci. **13**, 4981–4990 (2018)
2. S.P. Pattnaik, A. Behera, S. Martha, R. Acharya, K. Parida, J. Nanoparticle Res. **20**, 10 (2018)
3. S. Zarezadeh, A.H. Yangjeh, M. Mousavi, J. Photochem. Photobio. Chem. **379**, 11–23 (2019)
4. N. Mao1, Sci. Rep. **9**, 12383 (2019)
5. R. Cheng, X. Fan, M. Wang, M. Li, J. Tian, L. Zhang, R. Cheng, X. Fan, M. Wang, M. Li, J. Tian, L. Zhang, RSC Adv. **6**, 18990–18995 (2016)
6. J. Rashid, N. Parveen, T. Haq, A. Iqbal, S.H. Talib, S.U. Awan, N. Hussain, M. Zaheer, Chem. Cat. Chem. **10**, 5587–5592 (2018)
7. D.R. Paul1, S.P. Nehra, Environ. Sci. Poll. Res. **28**, 3888–3896 (2021)
8. L. Wanga, M. Zhang, J. Magn. Magn. Mater. **507**, 166841 (2020)
9. A.M. Gutierrez, T.D. Dziubla, J.Z. Hilt, Rev. Environ. Health. (2017). https://doi.org/10.1515/reveh-2016-0063
10. M.J. Jacinto, L.F. Ferreira, V.C. Silva, J. Sol-Gel Sci. Tech. **96**, 1–14 (2020)
11. N.P. Devi, M. Maisnam, Integr. Ferroelectr.133–141 (2020). https://doi.org/10.1080/10584587.2019.1674972
12. S. Kumar, S.T.B. Kumar, A. Baruah, V. Shanker, J. Phys. Chem. C **117**(49), 26135–26143(2013)
13. X. Jia, R. Dai, Y. Sun, H. Song, X. Wu, J. Mat. Sci.: Mat. Elect. **27**, 3791–3798 (2016)
14. D. Zhu, S. Liu, M. Chen, J. Zhang, X. Wang, Colloids Surf.: Physicochem. Eng. Aspects **537**, 372–382 (2018)
15. M. Mousavi, A.H. Yangjeh, D. Seifzadeh, K. Nakata, S. Vadivel, Adv. Powder Tech. **30**, 524–537 (2019)
16. J. Yang, H. Chen, J. Gao, T. Yan, F. Zhou, S. Cui, W. Bi, Mat. Lett. **164**, 183–189 (2016)
17. P. Liu, Q. Xu, D. Gao, S. Shi, B. Xia, J. Adv. Nanomater. **2** (2017)
18. Z. Zhu, Z. Lu, D. Wang, X. Tang, Y. Yan, W. Shi, Y. Wang, N. Gao, X. Yao, H. Dong, App. Catalysis B: Environ. **182**, 115–122 (2016)
19. M. Mousavi, A.H. Yangjeh, Mat. Chem. Phy. **163**, 421–430 (2015)
20. A.H. Yangjeh, A. Akhundi, J. Mol. Catalysis A: Chem. **415**, 122–130 (2016)
21. M. Mousavi, A.H. Yangjeh, Mat. Res. Bull. **105**, 159–171 (2018)
22. M. Mousavi. A.H. Yangjeh, J. Mat. Sci. **53**, 9046–9063 (2018)
23. A. Akhundi, A.H. Yangjeh, Mat. Sci. Semicond. Process. **39**, 162–171 (2015)
24. M. Mousavi, A.H. Yangjeh, J. Mat. Sci.: Mater. Electron. **27**, 8532–8545 (2016)
25. A. Akhundi, A.H. Yangjeh, Adv. Powder Tech. **29**, 94–105 (2018)
26. A. Kumara, A. Kumara, G. Sharmaa, A.H. Al-Muhtasebb, M. Naushadc, A.A. Ghfarc, F.J. Stadler, Chem. Eng. J. **334**, 462–478 (2018)
27. Z. Abbasia, A. Farrokhniaa, E.I. García-Lópezb, M. ZargarShoushtari, Phys. Chem. Res. **7**, 65–80 (2019)
28. A. H.-Yangjeh, M. Mousavi, Kazuya, J. Photochem. Photobiol. A: Chem. **368**, 120–136 (2019)
29. T. He1, Y. Wu, C. Jiang, Z. Chen, Y. Wang, G. Liu, Z. Xu, G. Ning, X. Chen, Y. Zhao, PLoS ONE. https://doi.org/10.1371/journal.pone.0237389

Chapter 26
Nanocomposites of NiO/Graphene as Efficient Electrocatalyst in Fuel Cell

Kashmiri Baruah and P. Deb

Abstract Ni systems have been considered a potential alternative of noble metal nanoparticles as electrocatalyst in fuel cell applications as it is low cost, possess good electrochemical stability, resistant to corrosion and exhibits electrocatalytic activity similar to the noble metal-based electrocatalysts. NiO supported on graphene nanosheets have been explored in varied fields like batteries/supercapacitors, sensors, electrocatalysts, etc. due to the synergistic effects of conductivity of graphene and electroactive sites of NiO. This work reviews the recent advancement in the application of NiO/graphene as electrocatalyst in fuel cells. The discussion surrounds on the role of surface and interface of the nanocomposite for imparting superior electrocatalytic property.

26.1 Introduction

Fuel cells (FCs) are attracting attention of researchers as a potential power source on account of higher power density and less creation of pollutants [1–3]. It basically converts the energy residing in chemical bond of the fuel into electrical energy [4, 5]. The electro-oxidation of the fuel is a sluggish reaction, thereby arising the necessity of an electrocatalyst to accelerate the redox reactions. Fuel crossover and catalyst poisoning by the intermediates formed are some of the disadvantages of FCs [6, 7]. By judiciously designing smart electrocatalyst with numerous electroactive sites and large surface area, one can efficiently increase the reaction kinetics and reduce the poisoning of the catalyst surface. Noble metal nanoparticles like Pt and Pd are considered hi-tech electrocatalyst till date. However, CO poisoning of these noble metal electrocatalysts reduces their efficiency and their scarcity on earth's crust compromise their practical implementation in large scale [7–9]. Alloying Pd/Pt with other transition metals (Sn, Ru, Ni, etc.) help reducing CO poisoning and the demand of Pt [10–12]. Transition metal oxides (TMO) are suitable replacement of the

K. Baruah · P. Deb (✉)
Department of Physics, Tezpur University (Central University), Tezpur 784028, India
e-mail: pdeb@tezu.ernet.in

R. G. Nair et al. (eds.), *Proceedings of 28th National Conference on Condensed Matter Physics*, Springer Proceedings in Physics 269,
https://doi.org/10.1007/978-981-16-5407-7_26

precious noble metals because of their natural abundance, non-toxicity, high stability, and brilliant electrochemical properties [13, 14]. Among different metal oxides, NiO and spinel oxides are grabbing attention [15]. In TMOs, the multiple oxidation states of the transition element cations offer reactive sites for electrooxidation and also result in different binding energy; moreover, the oxygen ions help in holding the charged entities on their surfaces without any blending [16–19]. Thus, TMOs are increasingly used in various applications like sensors [20], hydrogen production [21], energy storage [22], etc. It has been found by researchers that Ni is a potential competitor of noble metals as they possess good catalytic activity and are resistant to CO poisoning [23, 24]. NiO has attracted attention owing to its better electrocatalytic performance, abundance in nature, high stability and low cost [25]. NiO also shows excellent performance in other applications like lithium-ion batteries [26, 27], sensors [28], etc. However, poor conductivity of Ni-based materials increases the charge transfer resistance of the catalyst, hampering its electrocatalytic activity. This reduction in conductivity can be compensated by making nanocomposites of NiO with graphene, which has been used as support to many polymers and nanoparticles in multifarious applications, for instance electrocatalyst [29, 30], energy storage devices [31, 32], electrochemical sensors [33], etc. Thus, introducing graphene provides more active sites, which helps in preventing agglomeration of active material and hence making them more efficient and stable.

26.2 NiO/Graphene as Electrocatalyst in Fuel Cells

Nickel oxide decorated on graphene nanosheet (NiO/GNS) was devoted as a novel electrocatalyst for glucose oxidation reaction (GOR) [34]. The nanocomposite was synthesized by dissolving certain amount of sodium dodecyl sulfate in graphene oxide aqueous solution at room temperature. Nickel chloride hexahydrate was added into the solution with vigorous stirring until a homogeneous solution was obtained. After adding ethanol and urea into the solution, it was stirred for some time, followed by heat treatment in an autoclave at 160 °C for 10 h. Afterwards, the dried powder was annealed at 500 °C for 5 h to obtain NiO-graphene nanocomposite (NiO/GNS). Figure 26.1a depicts the X-Ray diffraction peaks of NiO-GNS and NiO. FESEM images of NiO in Fig. 26.1c indicates the formation of NiO nanoparticles which possess nanorod like morphology, which get agglomerated in the absence of GNS. While compositing with GNS (Fig. 26.1d–f), the NiO nanoparticles got dispersed uniformly on GNS surface, further preventing the agglomeration of NiO nanoparticles. BET isotherm of NiO-GNS exhibits type IV isotherm (Fig. 26.1b) which indicates the mesoporous nature of the nanocomposite with presence of slit like pores formed by aggregations of NiO nanoparticles. The nanocomposite NiO-GNS possess a surface area of 135.8 m^2/g and pore size of 14.8 nm. As compared with pristine NiO, the nanocomposite NiO/GNS exhibited two-fold higher GOR (Fig. 26.2a, b). This increased GOR is due to the improved electron transfer at the electrode interface and prevention of NiO aggregation due to the presence of graphene nanosheets.

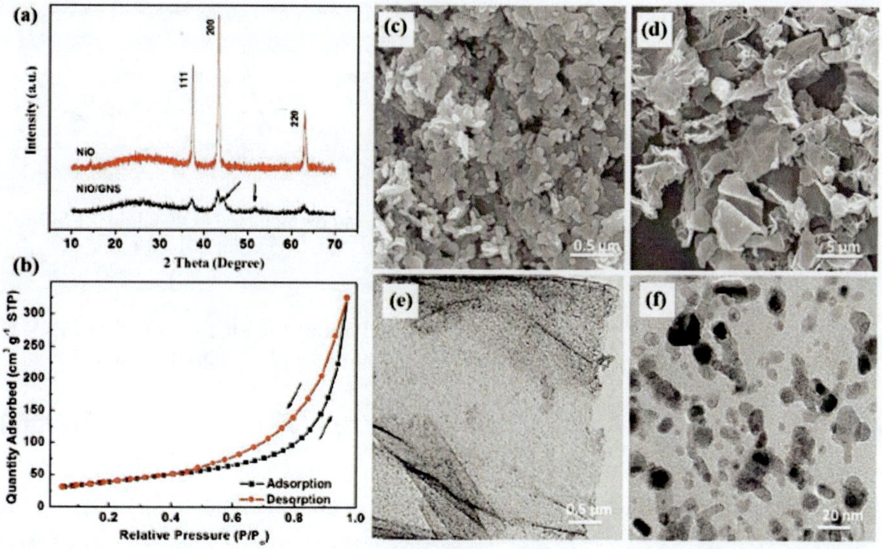

Fig. 26.1 a XRD patterns of NiO and NiO/GNS, b nitrogen adsorption–desorption isotherm of of NiO/GNS, c, d FESEM images of NiO and NiO/GNS, respectively. e, f TEM images of NiO/GNS with different magnification. Reproduced with permission under the terms of Creative Commons CC BY license, facilitated open access [34]. Copyright 2016 Springer Nature

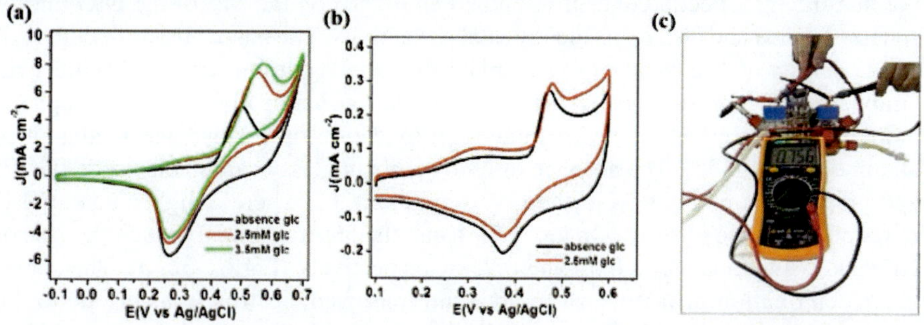

Fig. 26.2 a CV of NiO/GNS, b pristine NiO (both in 0.1 sodium hydroxide in presence and absence of glucose and scan rate 10 mV/s), and c Schematic of fuel cell test for measurement of open circuit voltage. Reproduced with permission under the terms of Creative Commons CC BY license, facilitated open access [34]. Copyright 2016 Springer Nature

The biofuel cell constructed using this nanocomposite as anode and Pt/C as cathode also showed high current density (0.66 A/m^2) and stability. The notable increase in anodic current density, while using NiO/GNS as the anode catalyst, implies its strong response towards GOR. Different states Ni^{2+} and Ni^{3+} participate in the redox reactions (explained in 26.1, 26.2); which is evident from the reduction of current density after addition of glucose (Fig. 26.2a), mainly due to the utilisation of Ni^{3+}

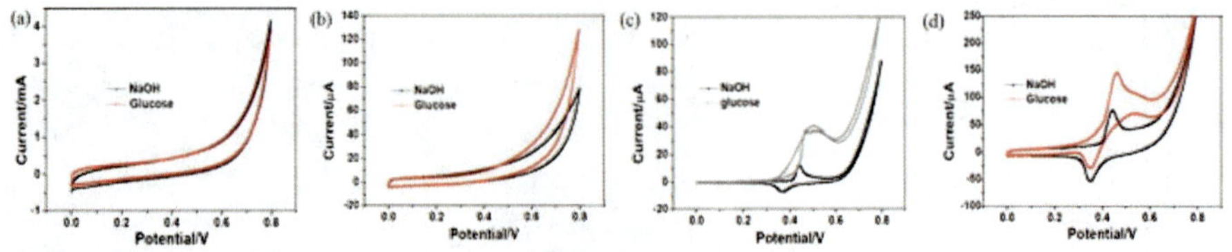

Fig. 26.3 CVs of **a** GCE, **b** graphene/GCE, **c** NiO/GCE, and **d** graphene/NiO/GCE in 0.1 M NaOH at a scan rate of 50 mVs^{-1}. Reproduced with permission [35]. Copyright 2014, Elsevier

by glucose oxidation. The redox reactions involved at the surface of NiO/GNS for glucose oxidation to form glucolactone are explained in 26.1–26.2:

$$NiO + OH^- \rightarrow NiOOH + e^- \quad (26.1)$$

$$NiOOH + glucose \rightarrow NiO + glucolactone \quad (26.2)$$

In redox reactions 26.1 and 26.2, the redox pair NiO/NiOOH is involved.

The glucose fuel cell test was performed and an open circuit voltage of 0.756 V was obtained (schematic shown in Fig. 26.2c). Another group developed a 3D porous NiO/graphene nanocomposite using biopolymer (pectin) template [35]. Pectin helped in the growth of NiO microspheres over graphene surface, hence forming the 3D porous structure. Pectin concentration had an impact on the size of the NiO microspheres formed and also on its electrocatalytic activity. The nanocomposite delivered power density of 3.63 Wm^{-2} in Microbial (Shewanella putrefaciens CN32) fuel cell. Graphene/NiO nanocomposite prepared by electrochemically reducing graphene oxide was assessed as an electrocatalyst in oxidation of glucose and methanol in alkaline medium [36]. The absence of redox pairs in glassy carbon electrode (GCE) and graphene is obvious from the CV (Fig. 26.3a, b), where a higher potential is required to oxidise glucose. On the other hand, the onset potential reduces in case of NiO and NiO/graphene to 0.32 and 0.25 V, respectively (Fig. 26.3c, d). The synergistic contribution of large surface area and improved electron transfer due to the development of electrical network of NiO nanoparticles decorated over the surfaces of graphene, leads to excellent electrocatalytic activity of the nanocomposite.

Hameed et al. [37] studied the effect of calcination temperature on the electrochemical performance (listed in Table 26.1) of the nanocomposite NiO/graphene

Table 26.1 Different electrochemical parameters of urea electrooxidation reaction at different NiO/Gr electrocatalysts. Reproduced with permission [37]. Copyright 2017, Elsevier

Electrocatalyst	E_{onset} (mV) (Ag/AgCl)	E_p (mV) (Ag/AgCl)	Ip (mA/cm^2)
NiO/Gr-200	386	710	30.94
NiO/Gr-300	365	718	11.30
NiO/Gr-400	349	630	8.93
NiO/Gr-500	297	567	8.36

obtained by coprecipitation method. Onset potential value of urea oxidation reaction (UOR) decreased as the calcination temperature increased. The lowest onset potential of NiO/Gr-500 (500 represents the calcination temperature) indicates its highest UOR activity. However, the better dispersion of NiO nanoparticles in NiO/Gr-200 leads to enhanced electrochemical performance towards UOR. Moreover, lower calcination temperature leads to lower charge transfer resistance. From this study, it can be understood that NiO nanoparticles calcined at lower temperature can provide better UOR performance in alkaline medium.

Pt/ NiO-GO nanocomposite grown on nickel foam when used as an anode catalyst for methanol oxidation reaction (MOR), exhibited better MOR activity and stability owing to the symbiotic effect of NiO-GO support and Pt nanoparticles [38].

26.3 Conclusion and Future Perspectives

In summary, various NiO/graphene nanocomposites have been studied and utilized as electrocatalyst in different fuel cells. The disadvantages of NiO nanoparticles like low conductivity and agglomeration have been overcome by constructing composite with graphene. The electrical network developed by the NiO nanoparticles decorated over graphene surfaces provides scope to the catalyst to enhance its catalytic performance and obtain an efficient and stable electrocatalyst. The excellent electrochemical performance of NiO/graphene nanocomposites provides scope in other electrochemical fields like lithium-ion batteries, supercapacitors, electrochemical sensors, etc.

Acknowledgements KB sincerely thanks UGC for providing the financial support by providing the UGC NFOBC fellowship.

References

1. D.J.L. Brett, A.R. Kucernak, P. Aguiar, S.C. Atkins, N.P. Brandon, R. Clague, L.F. Cohen, G. Hinds, C. Kalyvas, G.J. Offer, B. Ladewig, R. Maher, A. Marquis, P. Shearing, N. Vasileiadis, V. Vesovic, ChemPhysChem **11**, 2714 (2010)
2. S. Chu, A. Majumdar, Nature **488**, 294 (2012)
3. G. Che, B.B. Lakshmi, E.R. Fisher, C.R. Martin, Nature **393**, 346 (1998)
4. C. Rice, S. Ha, R.I. Masel, A. Wieckowski, J. Power Sour. **115**, 229 (2003)
5. M. Winter, R. Brodd, Chem. Rev. **104**, 4245 (2004)
6. D.D. James, P.G. Pickup, Electr. Acta **55**, 3824 (2010)
7. L. Yang, J. Ge, C. Liu, G. Wang, W. Xing, Curr. Opin. Electrochem. **4**, 83 (2017)
8. J.N. Tiwari, R.N. Tiwari, G. Singh, K.S. Kim, Nano Energy **2**, 553 (2013)
9. M.D. Bhatt, J.Y. Lee, Energy Fuels **34**, 6634 (2020)
10. M. Mansor, S.N. Timmiati, K.L. Lim, W.Y. Wong, S.K. Kamarudin, N.H. Nazirah Kamarudin, Int. J. Health Energy **44**, 14744 (2019)

11. J.C. Ng, C.Y. Tan, B.H. Ong, A. Matsuda, W.J. Basirun, W.K. Tan, R. Singh, B.K. Yap, Mater. Res. Bull. **112**, 213 (2019)
12. G.V. Reddy, P. Raghavendra, P.S. Chandana, L.S. Sarma, RSC Adv. **5**, 100522 (2015)
13. P. Manivasakan, P. Ramasamy, J. Kim, Nanoscale **6**, 9665 (2014)
14. Z. Yin, Y. Zheng, H. Wang, J. Li, Q. Zhu, Y. Wang, N. Ma, G. Hu, B. He, A. Knop-Gericke, R. Schlog, D. Ma, ACS Nano **11**, 12365 (2017)
15. H.S. Jadhav, S.M. Pawar, A.H. Jadhav, G.M. Thorat, J.G. Seo, Sci. Rep. **6**, 31120 (2016)
16. R.J. Lad, Phys. Struct. **185** (1996). https://doi.org/10.1016/s1573-4331(96)80010-2
17. D. Chimene, D.L. Alge, A.K. Gaharwar, Adv. Mater. **27**, 7261 (2015)
18. K. Kalantar-zadeh, J.Z. Ou, T. Daeneke, A. Mitchell, T. Sasaki, M.S. Fuhrer, Appl. Mater. Today **5**, 73 (2016)
19. M.C. Biesinger, B.P. Payne, A.P. Grosvenor, L.W. Lau, A.R. Gerson, R.S.C. Smart, Appl. Surf. Sci. **257**, 2717 (2011)
20. N. Joshi, T. Hayasaka, Y. Liu, H. Liu, O.N.O. Jr, L. Lin, Microchim. Acta **185**, 213 (2018)
21. A. Singhania, Catal. Lett. **148**, 1416 (2018)
22. K. Bhattacharya, P. Deb, Dalton Trans. **44**, 9221 (2015)
23. P. Alaba, C.H. Lee, F. Abnisa, M.K. Aroua, P. Cognet, Y. Peres, D. WMAW, Rev. Chem. Eng. (2020). https://doi.org/10.1515/revce-2019-0013
24. A. Ashok, A. Kumar, J. Ponraj, S.A. Mansour, F.J. Tarlochan, Electrochem. Soc. **165**, J3301 (2018)
25. S.-J. Li, W. Guo, B.-Q. Yuan, D.-J. Zhang, Z.-Q. Feng, J.-M. Du, Sens. Actuat. B: Chem. **240**, 398 (2017)
26. C. Xu, J. Sun, L. Gao, J. Power Sour. **196**, 5138 (2011)
27. X.H. Huang, J.P. Tu, B. Zhang, C.Q. Zhang, Y. Li, Y.F. Yuan, H.M. Wu, J. Power Sour. **161**, 541 (2006)
28. Q. Zhou, A. Umar, E.M. Sodk, A. Amine, L. Xu, Y. Gui, A.A. Ibrahim, R. Kumar, S. Baskoutas, Sens. Actuat. B: Chem. **259**, 604 (2018)
29. Y. Hu, J. Jin, P. Wu, H. Zhang, C. Cai, Electrochim. Acta **56**, 491 (2010)
30. W. He, H. Jiang, Y. Zhou, S. Yang, X. Xue, Z. Zou, X. Zhang, D.L. Akins, H. Yang, Carbon **50**, 265 (2012)
31. X. Hui, L. Qian, G. Harria, T. Wang, J. Che, Mater. Des. **109**, 242 (2016)
32. J. Lv, Z. Wang, H. Miura, Solid State Commun. **269**, 45 (2018)
33. Z. Yu, H. Li, X. Zhang, N. Liu, X. Zhang, Talanta **144**, 1 (2015)
34. G. Zeng, W. Li, S. Ci, J. Jia, Z. Wen, Sci. Rep. **6**, 36454 (2016). https://doi.org/10.1038/srep36454
35. X. Wu, Z. Shi, L. Zou, C.M. Li, Y. Qiao, J. Power Sour. **378**, 119 (2018)
36. S.-J. Li, N. Xia, X.-L. Lv, M.-M. Zhao, B.-Q.Yuan, H. Pang, Sens. Actuat. B: Chem **190**, 809 (2014)
37. R.M. Abdel Hameed, Shymaa S. Medany, J. Colloid Interface Sci. **508**, 291 (2017)
38. M.A. Kamyabi, H. Mohammadian, S. Jadali, M. Moharramnezhad, Electroanalysis **31**, 1501 (2010)

Chapter 27
A Review on Pure and Semiconductor Functionalized Ferroelectric Polymer-Based Memory Devices

Nipom Sekhar Das and Avijit Chowdhury

Abstract The bendable and transparent memory devices, comprising multifunctional and easily processable emerging ferroelectric polymeric materials, have attracted growing attention recently, especially in the branch of flexible neuromorphic electronics. Above all, the ferroelectric polymers like Poly(vinylidene fluoride) (PVDF) and its copolymer trifluoroethylene (TrFE) are very popular and well known for their excellent ferroelectric property, high compatibility with soft substrates, high chemical stability, and mechanical flexibility. Herein, the recent advances in pure and semiconductor functionalized PVDF, and its copolymer-based non-volatile memory devices are reviewed including their developments in flexible neuromorphic computing.

27.1 Introduction

Memory is essential for storing information and are the building block of digital electronics. Many modern technologies are exploited in designing flexible and transparent electronic and optoelectronic devices to overcome the theoretical and physical limits of conventional Si-based devices [1, 2].

The classification of electronic memory is schematically shown in Fig 27.1 and broadly divided into volatile and non-volatile memory. The non-volatile memory includes flexible flash memory [3], phase-change random access memory (PCRAM) [4], ferroelectric random-access memory (FeRAM), resistive random-access memory (RRAM) [5]. The RRAM is a modern electronic device with a simple Metal/ Insulator/Metal structure and has some desirable memory characteristics such as longer retention [6], excellent cycling stability [7], low operating voltage, high current on/off ratio, high bit density, etc. A schematic diagram of a RRAM device is shown in Fig 27.2. When a bias voltage is applied across the top and bottom electrode, distinct conduction states are found at elevated electric fields [8–10].

N. S. Das · A. Chowdhury (✉)
Organic Electronics and Sensor Laboratory, Department of Physics,
National Institute of Technology Silchar, Silchar, Assam 788010, India

R. G. Nair et al. (eds.), *Proceedings of 28th National Conference on Condensed Matter Physics*, Springer Proceedings in Physics 269,
https://doi.org/10.1007/978-981-16-5407-7_27

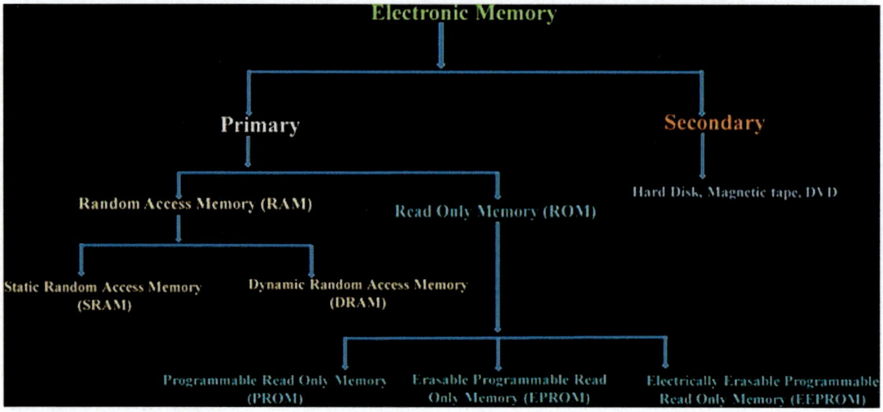

Fig. 27.1 Schematic diagram of classification of memory

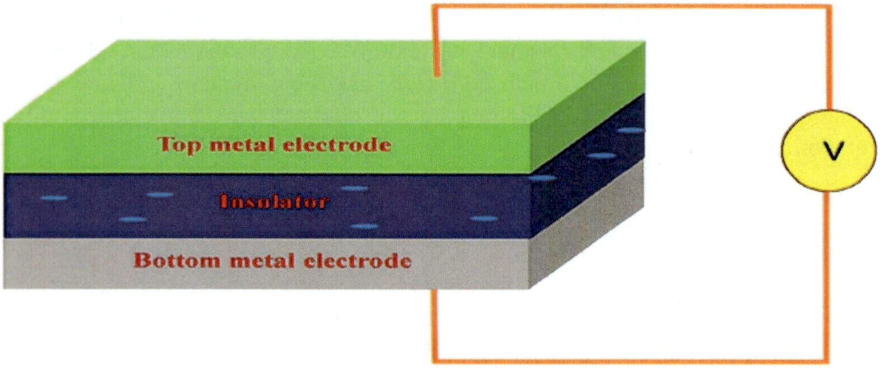

Fig. 27.2 Schematic diagram of resistive random-access memory

With the growing necessity in portable and wearable electronics, there is increasing attention on the flexible active layer fabricated onto a plastic substrate. The polymeric materials are highly stretchable and biocompatible, making them a promising candidate for flexible electronics [11]. Some of the polymeric materials like poly (methyl methacrylate) (PMMA), polystyrene (PS), Polyaniline (PANI), PVDF, etc. are found many applications in electronic devices. Alongside, the ferroelectric polymers possess inherent characteristics originated from their spontaneous and thermodynamically stable polarization states. Therefore, the ferroelectric polymers are the most attractive materials which can be integrated into flexible memory devices to improve the memory properties. Moreover, the resistive switching properties of the polymer film embedded with various organic/ inorganic nanofillers display improved performances resulting from the synergistic effect between the nanofillers and the polymers [12]. They exhibit unique characters in electronics as they display the advantages such as nontoxicity, low cost and lightweight characteristics. The research on ferroelectric polymeric materials is strongly inspired because of their tunable electric properties [13]. They are worth emerging as active material in non-volatile memory devices [14], organic solar cells [15], and piezoelectric sensors [16]. The PVDF and PVDF-TrFE based ferroelectric polymers have garnered growing attention in modern organic electronic devices recently. The ferroelectric memory can retain data even after the power is cycled. Thus they can be exploited as active material in non-volatile resistive memory devices, for example, PCRAM devices [4], flash memory devices, and RRAM devices. The ferroelectric polymers are broadly exploited in memory devices because of their polar phase [3] and many other advantages like fast switching, long durability and low power consumption. These features make them suitable for transparent and flexible data storage application [5, 17, 18]. Recently, the ferroelectric polymer-based flexible memory device is considered as an emerging material for the implementation of artificial synapses in neuromorphic computing because of their analog conduction response, fast operation speed, and low power consumption [19]. Due to their outstanding characteristics, the ferroelectric memory device is considered to replicate the artificial synapse [20–22].

27.2 Ferroelectric Polymeric Material and Its Application in Non-volatile Memory Device

The most recognized and familiar ferroelectric polymer is PVDF, having the monomer CH_2–CF_2. It is a semi-crystalline polymeric material and used in various research areas due to its pyroelectric and ferroelectric characteristics [23]. It can be found in different crystalline phases such as α, β, γ, and δ [24]. The α-PVDF is well known as the most stable phase but non-ferroelectric because of the absence of polarization. The β, γ, and δ phases are ferroelectric in nature, where β phase is straightforward to obtain, and also most of the memory devices use β-PVDF [23, 24]. The TrFE is used to obtain the ferroelectric properties of PVDF, where one hydrogen atom is replaced by fluorine to form P(VDF-TrFE) copolymer [25, 26]. It can be synthesized by using two homopolymers (PVDF and TrFE). The chemical structure of PVDF and P(VDF-TrFE) is schematically shown in Fig. 27.3.

The β-PVDF is highly polar, and thus their dipoles point in the same direction and exhibit permanent polarization, which renders its ferroelectric property. The ferroelectric polarization effect is introduced by the dipoles due to the difference in electronegativity between hydrogen (H) and fluorine (F) atoms. A schematic diagram of ferroelectric polarization effect of PVDF is shown in Fig. 27.4.

The PVDF is indeed very important due to its linear molecular structure, stable molecular dipole moment, and compact crystal structure [27]. Non-volatile memory devices comprising of organic ferroelectric polymers are a promising advancement toward the cost effective modern electronic technology [14] (Fig. 27.5).

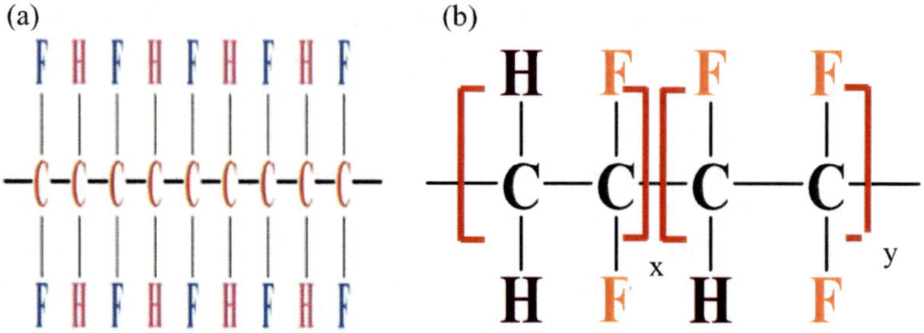

Fig. 27.3 Schematic diagram of **a** PVDF and **b** P(VDF-TrFE)

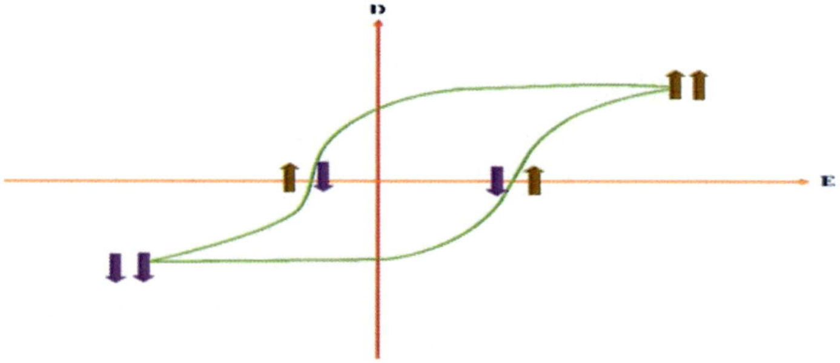

Fig. 27.4 Schematic diagram of ferroelectric effect shown by PVDF

The 2-D layered materials with exfoliated nanosheets are functionalized with ferroelectric polymers to improve the memory properties. The semiconductors functionalized with ferroelectric polymer improve the mechanical and electrical properties of the composite layer [28]. Some of the ferroelectric polymer and polymer nanocomposites-based memory devices are discussed in the Table 27.1.

27.3 Ferroelectric polymer for neuromorphic computing

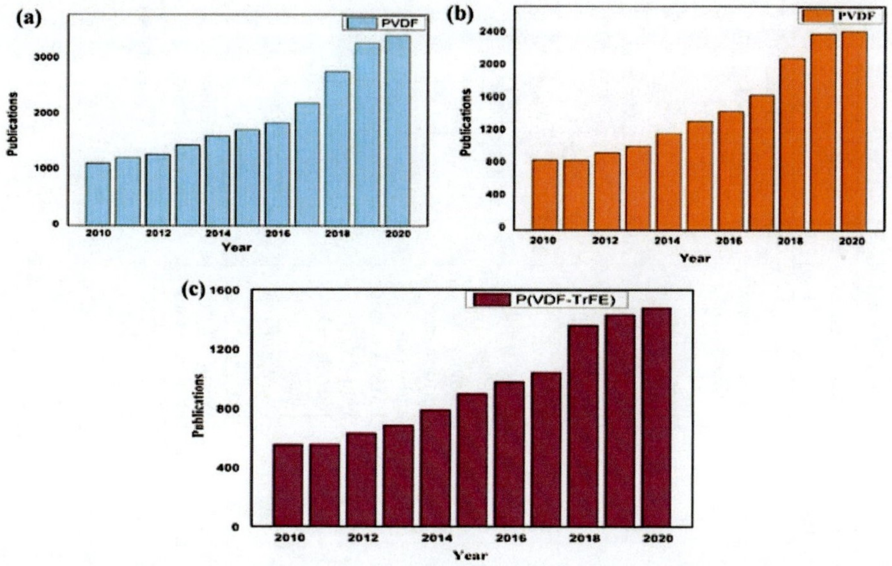

Fig. 27.5 Publication versus year graph of **a** Piezoelectric PVDF (Data obtained from Google Scholar from 2010 to 2020). **b** β-phase PVDF (Data obtained from Google Scholar from 2010 to 2020), **c** poly (vinylidene fluoride co trifluoroethylene) (P (VDF-TrFE)) (Data obtained from Google Scholar from 2010 to 2020)

The neural network based neuromorphic computing has attracted growing interest to process a large amount of unstructured data, unlike Von Neumann's computer architecture [36]. In neuromorphic computing, two neurons are connected via synapses. In recent years various neural network platforms have been carried through using synaptic devices based on ferroelectric transistors because of their fast switching and low power consumption properties [37]. Various synaptic functionalities such as potentiation, spike-time-dependent plasticity (STDP), can be implemented using the FRAM device. The ferroelectric device based on P(VDF-TrFE) is analogous to the biological synapse which is an emerging candidate in today's era.

27.4 Conclusions and Outlook

In this review, a comprehensive introduction of PVDF and its copolymers are highlighted, including the structure and performance of the ferroelectric polymers are briefly discussed. It can be concluded that ferroelectric polymer and its copolymers such as PVDF and P(VDF-TrFE) are promising polymeric materials for non-volatile resistive memory devices including neuromorphic computing. Alongside, the ferroelectric phases and analog conduction response of the PVDF has diverse applications in modern technology.

Table 27.1 Non-volatile resistive switching memory devices based on pure.and semiconductor functionalized ferroelectric polymer

Authors	Device structure	V_{set}(volt)	V_{reset}(volt)	I_{on}/I_{off}	Endurance cycles	Retention time (s)	Memory effect	Publication year	References
Hyuan et al.	Au/P(VDF-TrFE)/ITO	5.8	−6.1	10	10^3	–	Ferroelectric piezo response hysteresis loop and diode behavior	2016	Hyuan et al. [29]
Kim et al.	Au/P(VDF-TrFE) ZnO NPs/n^{++}Si	−3.90 to −2.74	2.81 to 2.60	2×10^7	10^4	10^4	Bipolar resistive switching	2018	Kim et al. [30]
Wang et al.	Ag/PVDF/PMMA/F8T2/Ag	–	–	10^3	–	10^3	The device shows good resistive memory performance and when PMMA doped with PVDF ferroelectric phase transformation occurs	2018	Wang et al. [31]
Khurana et al.	Pt/PVDF/GR/PVDF/ITO	2.2	−2.9	10^2	150	10^4	Worm characteristics	2014	Khurana et al. [32]

(continued)

Table 27.1 (continued)

Authors	Device structure	V_{set}(volt)	V_{reset}(volt)	I_{on}/I_{off}	Endurance cycles	Retention time (s)	Memory effect	Publication year	References
Xia et al.	Al/ P(VDF-TrFE)/PTFE/Al	3.4	−2.6	10^5	–	10^3	Exhibited the resistive switching property	2017	Xia et al. [33]
Lenz et al.	Au/PEDOT: PSS/P(VDF-TrFE)/PFO/Au	–	–	10^3	–	10^4	The memory diodes exhibited bistable electrical transport	2016	Lenz et al. [34]
Deepak et al.	Al/MoS_2/(PVDF-HFP)/ITO	2.74	−0.9	10^4	–	–	Bipolar resistive switching	2019	Deepak et al.[35]

References

1. M.A. Zidan, J.P. Strachan, Nat. Electron. **1**, 22 (2018)
2. J. Lee, W.D. Lu, Adv. Mater. **30**, 1702770 (2017)
3. S.-T. Han, Y. Zhou, Z.-X. Xu, L.-B. Huang, X.-B. Yang, V.A.L. Roy, Adv. Mater. **24**, 3556 (2012)
4. M. Wuttig, N. Yamada, Nat. Mater. **6**, 824 (2007)
5. T. Rueckes, K. Kim, E. Joselevich, G.Y. Tseng, C.-L. Cheung, C.M. Lieber, Science **289**, 94 (2000)
6. J.J. Yang, D.D. Strukov, D.R. Stewart, Nat. Nanotechnol. **8**, 13 (2013)
7. K.K. Kim, S.H. Jo, S. Gaba, W. Lu, Appl. Phys. Lett. **96**, 053106 (2010)
8. Y. Park, M.K. Kim, J.S. Lee, J. Mater. Chem. C **8**, 9163 (2020)
9. H.S.P. Wong, H.-Y. Lee, S. Yu, Y.-S. Chen, Y. Wu, P.S. Chen, B. Lee, F.T. Chen, M.-J. Tsai, Proc. IEEE **100**, 1951 (2012)
10. R. Waser, R. Dittmann, G. Staikov, K. Szot, Adv. Mater. **21**, 2632 (2009)
11. M. Amjadi, K.-U. Kyung, I. Park, M. Sitti, Adv. Funct. Mater. **26**, 1678 (2016)
12. C.W. Lin, T.S. Pan, M.C. Chen, Y.J. Yang, Y. Tai, Y.F. Chen, Appl. Phys. Lett. **99**, 023303 (2011)
13. X. Chen, X. Han, Q.-D. Shen, Adv. Electron. Mater. **3**, 1600460 (2017)
14. R.C.G. Naber, C. Tanase, P.W.M. Blom, G.H. Gelinck, A.W. Marsman, F.J. Touslager, S. Setayesh, D.M. de Leeuw, Nat. Mater. **4**, 243 (2005)
15. Y. Yuan, T.J. Reece, P. Sharma, S. Poddar, S. Ducharme, A. Gruverman, Y. Yang, J. Huang, Nat. Mater. **10**, 296 (2011)
16. X. Han, X. Chen, X. Tang, Y.-L. Chen, J.-H. Liu, Q.-D. Shen, Adv. Funct. Mater. **26**, 3640 (2016)
17. L. Ruan, D. Zhang, J. Tong, J. Kang, Y. Chang, L. Zhou, G. Qin, X. Zhang, Ferroelectr. Appl. (2018)
18. J.F. Scott, Science **315**, 954 (2007)
19. S. Yu, H.-Y. Chen, B. Gao, J. Kang, H.-S.P. Wong, ACS Nano **7**, 2320 (2013)
20. S. Boyn et al., Nat. Commun. **8**, 14736 (2017)
21. C. Yoon, J.H. Lee, S. Lee, J.H. Jeon, J.T. Jang, D.H. Kim, Y.H. Kim, B.H. Park, Nano Lett. **17**, 1949 (2017)
22. A. Chanthbouala et al., Nat. Mater. **11**, 860 (2012)
23. H. Kawai, Jpn. J. Appl. Phys. **8**, 975 (1965)
24. H. Li, R. Wang, S.-T. Han, Y. Zhou, Polym. Int. **69** (2020)
25. H. Zhu, C. Fu, M. Mitsuishi, Polym. Int. (2020)
26. T. Lenz, H.S. Dehsari, K. Asadi, P.W.M. Blom, W.A. Groen, D.M. de Leeuw, Appl. Phys. Lett. **109**, 133302 (2016)
27. M. Poulsen, S. Ducharme, IEEE Trans. Dielectr. Electr. Insul. **17**, 1028 (2010)
28. Q. Jia, X. Huang, G. Wang, J. Diao, P. Jiang, J. Phys. Chem. C **120**, 10206 (2016)
29. S. Hyun, O. Kwon, C. Choi, K.L. Vincent Joseph, Y. Kim, J.K. Kim, ACS. Appl. Mater. Interfaces **8**, 27074 (2016)
30. T.Y. Kim, G. Anoop, Y.J. Son, S.H. Kim, E. Lee, J.Y. Jo, Phys. Chem. Chem. Phys. **20**, 16176 (2018)
31. H. Wang, C.-T. Wang, F. Xu, J. Yang, J. Liu, W. Cai, G. Zhu, Polymer **153** (2018)
32. G. Khurana, P. Misra, R.S. Katiyar, Carbon **76**, 341 (2014)
33. W. Xia, C. Peter, J. Weng, J. Zhang, H. Kliem, Y. Jiang, G. Zhu, A.C.S. Appl, Mater. Interfaces **9**, 12130 (2017)
34. T. Lenz et al., Adv. Funct. Mater. **26**, 5111 (2016)
35. D. Deepak, R. Deb, M.G. Nair, S. Halder, A.L. Sharma, S.R. Mohapatra, Mater. Today: Proc. **18**, 5447 (2019)
36. K. Roy, A. Jaiswal, P. Panda, Nature **575**, 607 (2019)
37. Z. Yin, B. Tian, Q. Zhu, C. Duan, Polymers **11**, 2033 (2019)